John O'Sullivan

European Missions to the International Space Station

2013 to 2019

 Springer

Published in association with
Praxis Publishing
Chichester, UK

John O'Sullivan
County Cork, Ireland

SPRINGER-PRAXIS BOOKS IN SPACE EXPLORATION

Springer Praxis Books
Space Exploration
ISBN 978-3-030-30325-9   ISBN 978-3-030-30326-6   (eBook)
https://doi.org/10.1007/978-3-030-30326-6

Cover design: Jim Wilkie

This Springer imprint is published by the registered company Springer Nature Switzerland AG
The registered company address is: Gewerbestrasse 11, 6330 Cham, Switzerland

To Kate and Victor

Contents

Foreword

The general public initially associated astronauts with the United States and cosmonauts with the then-USSR. On the basis of bilateral agreements, several European countries managed to have European citizens fly on a number of space missions, be it with NASA (especially involving the provision of the ESA Spacelab module in the shuttle cargo bay) or with the Soyuz rocket to the Russian MIR station.

Although this was a highly interesting experience, it put European astronauts on the same footing as many other international astronauts who flew with those two major space powers, typically as payload specialists and often chosen based upon political motivations. There is no doubt that such flights were very important for the image of space in the respective European countries, but it did not provide the opportunity to develop a solid European astronaut program.

This changed near the end of the 1990s, when Europe started its own space transportation project, called Hermes. This small spacecraft was scheduled to fly from Kourou with European astronauts onboard. As a consequence, it was decided to initiate a number of astronaut training facilities with a dedicated European astronaut headquarters in Cologne, on the premises of the German space agency, DLR. This astronaut center was named the European Astronaut Centre (EAC). As a member of the Hermes management team, I was detached in 1989 to the forerunner of this center, responsible for coordination of European astronauts.

As a next step, it was decided to group the existing astronauts as well as a number of newly recruited ones in a joint European Astronaut Corps. On the basis of the estimated number of flight opportunities a first corps of twelve astronauts was formed, a majority of them experienced test pilots, in view of the planned Hermes flights.

Consequently, astronauts received basic training at EAC and were then integrated into the astronaut classes in NASA as mission specialists, or assigned to missions on board of the Russian Mir station, in particular Euromir94 and Euromir95, where I had the privilege of coordinating the European astronaut activities. Both missions were long-duration flights and a first exposure to the very rich experience of Russia with this type of flight. Several other flights followed, also national ones, all in preparation for the flights that would go to the International Space Station (ISS).

Fig. F.1. European Astronaut Centre (EAC), Cologne, Germany. (Image courtesy of ESA)

Fig. F.2. Walter Peeters (Wikipedia)

It was these steps that led towards the recognition of European astronauts as a full partner in the ISS era, with several flights as described in this excellent book. Books of this
nature are important, as active participants in the activities are caught in recurring cycles
of flight preparation and operations, in combination with chosen experiments and training.
Unfortunately, leaves little time for participants to document all their experiences and log
them for the future. Books of this kind, therefore, are a treasure for future researchers
interested in the evolution of European astronaut activities.

Prof. dr. W. Peeters, President (Emeritus), International Space University, Strasbourg,
France, and former head of Coordination, European Astronaut Centre, Cologne, Germany

Preface

I wrote my first book, *In the Footsteps of Columbus, European Missions to the International Space Station,* because I wanted to read it. I am a reader and collector of the Outward Odyssey series of books published by the University of Nebraska Press and edited by Colin Burgess. In April 2012 Colin posted on the collectSPACE website that UNP had asked him to explore the possibility of going beyond the 12 books in the series. He was asking readers for suggestions. My suggestions were a book on the Soyuz/Salyut/Mir missions and a book on the ESA missions to Mir, the ISS and shuttles. Colin's reply was that the UNP didn't see a market for non-U.S. stories. Demonstrating the depth of the forum users, David J. Shayler posted a reply mentioning Clive Horwood of Praxis. I contacted Clive proposing a book covering the European missions to the ISS… and the rest, as they say, is history.

I am extremely grateful to Clive and Maury Solomon of Springer for giving me the opportunity to continue telling the story of European spaceflight with this book.[1]

The history of European human spaceflight is not as straightforward as its American or Russian endeavors. Europe wasn't a competitor in the 'space race.' As a collection of nations with different languages, cultures and goals, the vision for space has been complex. For the first three decades of the space age, Europe was divided by the Iron Curtain. Even today, the European Space Agency does not build or fly a human-rated spacecraft. But despite all these factors there is a rich history of Europeans traveling to space on a variety of spacecraft and performing a variety of missions.[2]

As Europe isn't a single country with a human space program, European citizens must 'hitch a ride' to get into space. This has resulted in many different routes to orbit. Before the period covered by these books, astronauts from communist countries and from France had flown on Soviet Soyuz spacecraft to the Salyut and Mir space stations. Later,

---

[1] And of course, the opportunity to write my second book, *Japanese Missions to the International Space Station, Hope from the East.*

[2] In December 2013, NASA and ESA agreed that the European Service Module (ESM), based on the Automated Transfer Vehicle (ATV), would provide power and propulsion for the first Orion mission, *Artemis 1* (formerly Exploration Mission 1). *Artemis 2* is planned to be a crewed mission.

astronauts from other Western European space agencies and ESA flew to Mir. Western Europeans represented their national space agencies and ESA by flying on NASA space shuttle missions. Naturalized U.S. citizens from around the world, including quite a few Europeans, succeeded in joining NASA's astronaut corps by applying to the Johnson Space Center in Houston, Texas. And finally, almost as a footnote, there have been several European-born American astronauts.

Within the scope of these books are the European ESA astronauts who have flown to the ISS. Even this story isn't straightforward, because although most flew as members of the ESA astronaut corps, one flew as a French CNES astronaut prior to joining ESA, and others flew as representatives of their national space agencies and not as ESA astronauts.

In my previous book, *In the Footsteps of Columbus*, as a prelude to detailing the European missions to the ISS, I wrote a brief history of European human spaceflight, before the ISS. It sets the scene for the European missions in that book and this one. It was written in chronological order and contained flights of the following types:

- Intercosmos missions: Eastern Europeans on Soviet missions to Salyut 6. For simplicity, including Bulgarian Alexandrov's mission to Mir.
- ESA shuttle missions: ESA astronauts on space shuttle missions, including missions that carried the Spacelab module.
- Non-ESA shuttle missions: CNES/DLR/ASI astronauts on space shuttle missions. Some may or may not have flown as ESA astronauts on other missions. These include payload specialists not assigned by a space agency, e.g., Dirk Frimout.
- ESA astronauts on Soyuz missions.
- CNES/ASA/DLR astronauts on Soyuz missions.
- Miscellaneous: Helen Sharman was selected to fly to Mir on a mission funded by private UK companies without the assistance of either the UK government or ESA.

However, I have considered the following technically European human spaceflights to lie *outside* the scope of this book:

- Cosmonauts: For practical reasons, I have excluded all cosmonauts from European Russia (west of the Urals), Ukraine, Belorussia and the Baltic countries.
- Naturalized and dual citizens: Europeans who gained citizenship of USA/Canada and joined NASA/CSA, or who flew as payload specialists on the shuttle. For example:
  - Lodewijk van den Berg (Netherlands born U.S. citizen), who flew as payload specialist on STS-51B.
  - Michael Foale (UK born, dual UK/U.S. citizen), who has flown to space six times on both Soyuz and shuttle spacecraft and has lived on Mir and commanded the ISS.
  - Michael Lopez-Alegria (Spanish born U.S. citizen), who has flown in space four times on both Soyuz and shuttle spacecraft and has commanded the ISS.
  - Bjarni Trygvasson (Iceland born Canadian citizen), who flew as a payload specialist on STS-85 representing CSA.
  - Piers Sellers (UK born U.S. citizen), who has flown on three Shuttle missions as mission specialist and visited the ISS.

- Nicholas Patrick (UK born U.S. citizen) who has flown on two shuttle missions as mission specialist and visited the ISS.
- Charles Simonyi (Hungarian born U.S. citizen), who has flown twice on Soyuz missions to the ISS as a 'space tourist.'

- U.S. citizens born abroad to U.S. citizen parents:

- Michael Collins (born in Italy), who flew on Gemini 10 and Apollo 11. He is undeniably the first European-born astronaut, but for consistency has to be excluded on the grounds of his American parentage and citizenship.[3]
- Gregory Johnson (born in the UK), who flew on two shuttle missions as pilot and visited the ISS.
- Richard Garriott (born in the UK), who flew on Soyuz TMA-31 to the ISS as a 'space tourist' and second-generation spacefarer.

I have used a variety of sources for the material in this book, including NASA and ESA mission reports, astronaut biographies and blogs, contemporary magazines and reference websites such as the encyclopedic www.spacefacts.de. As a result, the content of individual chapters may differ in tone or focus from the others.

If blog entries by astronauts are relatively brief, I put them in the appropriate place in the chronology. For other longer FAQs and blog entries that cover broader issues, I put them in the appendices.

For mission names, I decided not to use the uppercase 'ISS,' which ESA insisted on shoehorning into their titles. So, for example, I refer to Odissea rather than OdISSea, because the latter seems to be distracting to the reader.

There is a rich story of human spaceflight happening between and around the European missions detailed in this book. Although I have endeavored to inform the reader of key missions and events in American and Soviet/Russian spaceflight during the time period, these missions lie outside the scope of this book; they are, however, well covered in other Springer-Praxis publications.

When selecting terminology, I've used the term astronaut when describing flights on American spacecraft and cosmonaut for flights on Soviet or Russian spacecraft. The same spacefarer could be described as both over the course of the book, as many Europeans have flown on both American and Russian craft. As yet there have been no European taikonauts…

The predecessor of this book, published in 2016, covered the missions from Umberto Guidoni's STS-100 mission in 2001 to André Kuipers' Promisse mission of 2011. These are briefly reviewed in the chapter 1 of this book. This book covers the missions from Luca Parmitano's Volare mission to his 2019 Beyond mission. At the time of submitting this manuscript for publication, Luca Parmitano had just launched on his Beyond mission to the ISS to join Expedition 60 and will go onto command Expedition 61. The first weeks of this mission are covered here.

This book covers a period of operational stability on board the station. The construction phase is complete and the space shuttle is retired. As a result, the later missions, with the exception of Andreas Mogensen's Iriss mission, were all full six-month expeditions.

---

[3] A good tip when setting quiz questions!

While the first volume covered 18 missions and 918 days in space, this new volume covers only 8 additional missions but over 1,200 additional days in space. For the Volare and Blue Dot missions, I continued the same approach as for the previous long duration mission chapters. However, for the later chapters, to keep page count down, I followed the same pattern as I had for my book *Japanese Missions to the International Space Station, Hope from the East*, i.e., I avoided repetitive coverage of the mundane station tasks, such as routine physical exercise, medical tests, station maintenance (there is a lot of water and waste to filter, transfer and reuse, etc.), housekeeping, eating meals and logging food intake, charging and changing batteries and packing and unpacking cargo. I also focused on the activities of the ESA astronauts while minimizing coverage of the work carried out by their American and Russian colleagues.

It is hoped that the ISS remains in orbit for many years to come and that European astronauts continue to live and work on the station performing maintenance, scientific research and educational outreach. If that happens, we can continue to chronicle their achievements in the future.

John O'Sullivan
September 2019

Acknowledgements

I must thank Clive Horwood of Praxis in England and Maury Solomon of Springer in New York for allowing me to continue telling this story of human spaceflight. I would like to thank Maury and Hannah Kaufman for editing the manuscript. I would like to thank Maury and Hannah Kaufman for editing the manuscript and David M. Harland for assisting with the layout.

I would like to thank Jim Wilkie for creating a stunning cover for the third time of asking. And finally, I would like to thank Prof. Walter Peeters of the ISU for contributing the foreword to the book.

About the Author

John O'Sullivan MSc, BE, BSc (Hons), Dip PM, CEng FIEI, PMP, FSP, CMSE® has over 25 years' experience in the automation and control sector delivering solutions to the pharmaceutical and medical device industry in Ireland. He is the Engineering Director of Douglas Control and Automation in Cork, Ireland.

He has an engineering degree in Electrical Engineering from University College Cork, a science degree in Astronomy and Planetary Science from the Open University and a Diploma in Project Management from the Cork Institute of Technology. He is a Chartered Engineer and Fellow of Engineers Ireland and a Project Management Professional with the Project Management Institute. He is a Functional Safety Professional and Certified Machine Safety Expert, both certified by TÜV. He is, at the time of publication, studying management at the Michael Smurfit Graduate Business School at University College Dublin.

He has always had a fascination with aviation and space, leading him to gain his Private Pilot License in 2003 and to study astronomy.

He lectures in Automation at CIT in Cork and has written articles for Ireland's *Engineers* journal and British Interplanetary Society's *Spaceflight* magazine.

He was an unsuccessful candidate for the ESA Astronaut Corps in 2008. He lives in East Cork, Ireland.

His first book, *In the Footsteps of Columbus,* covering early European missions to the International Space Station, was published by Springer-Praxis in 2016. His second book, *Japanese Missions to the International Space Station, Hope from the East,* was published in 2018. This is his third book.

Acronyms

3AF	French Aeronautics and Astronautics Association
AAS	American Astronautical Society
ABC	American Broadcasting Company
ABRS	Advanced Biological Research System
ACBM	Active Common Berthing Mechanism
ACE	Advanced Colloids Experiment
ACES	Atomic Clock Ensemble in Space
AGU	American Geophysical Union
AIAA	American Institute of Aeronautics and Astronautics
AMS	Alpha Magnetic Spectrometer
APEX	Advanced Plant Experiments
ARED	Advanced Resistive Exercise Device
ARISS	Amateur Radio on the ISS
ASE	Association of Space Explorers
ASI	Agenzia Spaziale Italiana (Italian Space Agency)
ASTP	Apollo Soyuz Test Programme
ATV	Automated Transfer Vehicle
BASS	Burning and Suppression of Solids
BBC	British Broadcasting Corporation
BCAT	Binary Colloidal Alloy Test
BCAT-KP	Binary Colloidal Alloy Test Low Gravity Phase Kinetics-Critical Point
BE	Bachelor of Engineering Degree
BEAM	Bigelow Expandable Activity Module
BRIC	Biological Research in Canisters
BS	Bachelor of Science Degree
BSc	Bachelor of Science Degree
C2V2	Common Communications for Visiting Vehicles
CAL	Cold Atom Lab
CATS	Cloud Aerosol Transport System

CAVES	Cooperative Adventure for Valuing and Exercising human behaviour and performance Skills
CBCS	Centreline Berthing Camera System
CBM	Common Berthing Mechanism
CDR	Commander
CDRA	Carbon Dioxide Removal Assembly
CDT	Central Daylight Time
CEng	Chartered Engineer
CEO	Chief Operating Officer
CET	Central European Time
CEVIS	Cycle Ergometer with Vibration Isolation and Stabilisation
CIMON	Crew Interactive Mobile CompanioN
CIR	Combustion Integrated Rack
CIT	Cork Institute of Technology
CMG	Companion of the Order of St. Michael and St. George
CMO	Crew Medical Officer
CMSE	Certified Machine Safety Expert
CNES	Centre National d'Études Spatiale (Centre for Space Studies)
CNN	Cable News Network
COLBERT	Combined Operational Load Bearing External Resistance Treadmill
COSTAR	Corrective Optics for Space Telescope Axial Replacement
COTS	Commercial Orbital Transportation Services
CREAM	Cosmic Ray Energetics and Mass Investigation
CRS	Commercial Resupply Services
CSA	Canadian Space Agency
CSLM	Coarsening in Solid Liquid Mixtures
C-SPAN	Cable-Satellite Public Affairs Network
CST	Central Standard Time
CUCU	COTS UHF Communications Unit
DAMA	'Dark Matter'
DC	District of Columbia
DELIC	Device for the study of Critical Liquids and Crystallisation
DELTA	Dutch Expedition for Life Science, Technology and Atmospheric Research
DFG	Deutsche Forschungsgemeinschaft (German Research Foundation)
DGG	Deutsche Geophysikalische Gesellschaft (German Geophysical Society)
DLR	Deutsches Zentrum für Luft- und Raumfahrt (German Centre for Flight and Space Flight) previously DVLR
DNA	Deoxyribonucleic acid
DSLR	Digital Single-Lens Reflex
DTU	Danmarks Tekniske Universitet (Technical University of Denmark)
DVD	Digital Versatile Disc
EAC	European Astronaut Centre
ECLSS	Environment Control & Life Support Systems

ECOSTRESS	ECOsystem Spaceborne Thermal Radiometer Experiment on Space Station
EDR	European Drawer Rack
EDT	Eastern Daylight Time
EGU	European Geosciences Union
ELF	Electrostatic Levitation Furnace
EML	Electromagnetic Levitator
EMU	Extravehicular Mobility Unit
EPNR	École du Personnel Navigant D'essais et de Reception
ESA	European Space Agency
ESERO	European Space Education Resource Office
ESPN	Entertainment and Sports Programming Network
EST	Eastern Standard Time
ESTEC	European Space Research and Technology Centre
ETPS	Empire Test Pilot School
ETVCG	External Television Camera Group
EVA	Extravehicular Activity
EWO	Electronics Warfare Officer
FAQ	Frequently Asked Questions
FASES	Fundamental and Applied Studies of Emulsion Stability
FASTER	Facility for Absorption and Surface Tension
FGB	Functional Cargo Block
FIEI	Fellow of the Institution of Engineers of Ireland
FIU	Florida International University
FLEX	Flame Extinguishment Experiment
FPEF	Fluid Physics Experiment Facility
FRAM	Flight Releasable Attachment Mechanism
FSP	Functional Safety Professional
GLACIER	General Laboratory Active Cryogenic ISS Experiment Refrigerator
GMT	Greenwich Mean Time
GPS	Global Positioning System
HDEV	High Definition Earth-Viewing
HIV	Human Immunodeficiency Virus
HTV	H-II Transfer Vehicle
IAVCEI	International Association of Volcanology and Chemistry of Earth's Interior
IBM	International Business Machines
ICE	International Commercial Experiments
IDA	International Docking Adapter
IMAX	Image MAXimum
ISAE	Institute Superieure de l'Aeronautique et de l'Espace
ISERV	ISS SERVIR Environmental Research and Visualization System
ISS	International Space Station
ISU	International Space University
ITV	Independent Television

JAXA	Japanese Aerospace Exploration Agency
JCO	Junior Commissioned Officer
JEF	Japanese Exposed Facility
JEM	Japanese Experiment Module
JEMRMS	Japanese Experiment Module Remote Manipulator System
JSC	Johnson Space Center
KTH	Kungliga Tekniska Högskolan
LBNP	Lower Body Negative Pressure
LED	Light-Emitting Diode
LEE	Latching End Effector
LEO	Low Earth Orbit
LF	Logistics flight
LMM	Light Microscopy Module
LMU	Ludwig Maximilians University
MARES	Muscle Atrophy Resistive Exercise
MBSU	Main Bus Switching Unit
MCA	Major Constituent Analyser
MCOP	Multilateral Crew Operations Panel
MDM	Multiplexer-Demultiplexer
MED	Miniature Exercise Device
MELFI	Minus-Eighty Laboratory Freezer for ISS
MELiSSA	Micro-Ecological Life Support System Alternative
MEP	Member of European Parliament
MERLIN	Microgravity Experiment Research Locker Incubator
METERON	Multi-Purpose End-to-End Robotic Operations Network
MISSE	Materials on International Space Station Experiment
MLM	Multi-purpose Laboratory Module
MPLM	Multi-Purpose Logistics Module
MSc	Master of Science Degree
MSG	Microgravity Science Glovebox
MSPR	Multipurpose Small Payload Rack
MT	Mobile Transporter
MTRA	Mobile Transport Relay Assembly
NASA	National Aeronautics and Space Administration
NATO	North Atlantic Treaty Alliance
NBC	National Broadcasting Company
NEEMO	NASA Extreme Environment Mission Operations
NFL	National Football League
NORS	Nitrogen Oxygen Recharge System
OCT	Optical Coherence Tomography
ORMatE	Optical Reflector Materials Experiment
ORU	Orbital Replacement Unit
OSTPV	Onboard Short-Term Planner View
OU	Open University
PCBM	Passive Common Berthing Mechanism

PCM	Pressurized Cargo Module
PDAM	Pre-Determined Avoidance Manoeuvre
PhD	Doctor of Philosophy
PLC	Pressurized Logistics Carrier
PMA	Pressurized Mating Adaptor
PMM	Permanent Multi-purpose Module
PMP	Project Management Professional
PLT	Pilot
PPL	Private Pilot License
RAF	Royal Air Force
RGB	Radiator Grapple Bars
RINGS	Resonant Inductive Near-field Generation System
RKA	Russian Federal Space Agency
ROBoT	Robotics OnBoard Trainer
RPCM	Remote Power Controller Module
RRM	Robotic Refueling Mission
RS	Russian Segment
RTL	Radio Télévision Luxembourg
SAFER	Simplified Aid for EVA Rescue
SAMS	Space Acceleration Measurement System
SAW	Solar Array Wing
SCU	Service and Cooling Umbilical
SEAL	Sea, Air and Land
SLAMMD	Space Linear Acceleration Mass Measurement Device
SLC	Space Launch Complex
SM	Service Module
SPDM	Special Purpose Dexterous Manipulator
SPHERES	Synchronised Position, Hold, Engage, Reorient, Experimental Satellites
SSRMS	Space Station Remote Manipulator System
SSU	Sequential Shunt Unit
STEM	Science, Technology, Engineering and Mathematics
STP	Space Test Programme
STS	Space Transportation System
SVE	Société Volcanologique Européenne (European Volconological Society)
TLA	Three Letter Acronym
TMA	Transport Modified Anthropometric
TORU	Telerobotically Operated Rendezvous System
TÜV	Technischer Überwachungsverein (Technical Inspection Association)
TVIS	Treadmill with Vibration Isolation System
UF	Utilization Flight
UHF	Ultra High Frequency
UK	United Kingdom
UKSA	United Kingdom Space Agency
ULA	United Launch Alliance
ULF	Utilization and Logistics Flight

UNP	University of Nebraska Press
UPLC	Unpressurized Logistics Carrier
USAFE	United States Air Force Europe
USOC	User Support and Operation Centre
USOS	United States Orbital Segment
UTA	Utility Transfer Assembly
UTC	Universal Time Coordinate / Coordinated Universal Time
VIABLE	eVatuatIon And monitoring of microBiofiLms insiDe the ISS
WETA	Wireless Video System External Transceiver Assembly
WORF	Window Observational Research Facility
WPA	Water Processor Assembly
WRS	Water Recovery System
WHC	Waste and Hygiene Compartment

Part I

Background

1

Earlier European Missions to the ISS

This introduction briefly covers the European missions to the International Space Station up to 2012 after which this book takes up the story. These earlier missions are covered in detail in the book *In the Footsteps of Columbus: European Missions to the International Space Station,* by the same author.

On April 23, 2001, Roman physicist Umberto Guidoni crossed from the space shuttle orbiter Endeavour into the International Space Station. In doing so, he became the first European to enter the station. This was his second and final space-flight. The mission was literally instrumental in the construction of the station as it delivered the Canadarm2 robot arm, the main component of the Space Station Remote Manipulator System (SSRMS). Endeavour also carried the *Raffaello* Multi-Purpose Logistics Module (MPLM) which was temporarily attached to the station in order to deliver supplies and equipment to the crew. This was the second MPLM to travel to the station and the first time *Raffaello* made the trip. Between *Raffaello* and *Leonardo* (*Donatello* never flew), they replenished the ISS twelve times. *Leonardo* was eventually reconfigured to remain at the station as the redes-ignated Permanent Multipurpose Module. Guidoni returned to Earth for the last time on May 1 but he kick started a sequence of European missions that would help construct the largest spacecraft ever built. Once construction was complete, Europeans would live and work on board the ISS for decades. Guidoni went on to become a Member of the European Parliament (MEP).

Claudie Haigneré arrived at the station in a very different manner a few months later. The French woman, on her second spaceflight, arrived on board the Russian Soyuz TM-33 capsule, docking on October 23, 2001. At the time, that model of Soyuz did not have the capacity to remain docked at the ISS for the six-month duration of an Expedition. For that reason, long duration crews did not return to Earth in the same spacecraft in which they launched. A short duration visiting crew would deliver the replacement. These swaps were timed so that two

© Springer Nature Switzerland AG 2020
J. O'Sullivan, *European Missions to the International Space Station,*
Springer Praxis Books, https://doi.org/10.1007/978-3-030-30326-6_1

Expedition members would be replaced each time. As there was a third seat in the Soyuz, a guest could visit the station for a week or so, arriving with the new crew and returning with the old. Or, as in this case, a specific crew would launch and return in a week. Haigneré's Andromède mission was one such "visiting mission".[1] She returned to Earth for the final time on Soyuz TM-32 and went on to be a Minister in the French government as well as an advisor to the ESA director general.

Italian test pilot, Roberto Vittori took advantage of the lifeboat swap and Italy's unique position in ESA[2] to visit the station in late April 2002. He arrived on Soyuz TM-34 and returned to Earth on Soyuz TM-33. He was accompanied on his Marco Polo mission by the second "space tourist", Mark Shuttleworth of South Africa, the first African in space. Vittori would visit the station on two more occasions and, after his last flight, work with ASI on the technical scientific committee and later as the Space Attaché at the Embassy of Italy and head of the ASI Office in Washington DC.

Phillipe Perrin is an outlier in the story of European missions to the ISS. Perrin was a member of the French space agency Centre National d'Études Spatiales (CNES) and was seconded to NASA and trained in Houston to be a NASA mission specialist. It was in this capacity that he visited the station as a crewmember of STS-111 in June 2002. He would later join the ESA astronaut corps after his mission but never flew in space as an ESA astronaut. So, he is the only European astronaut to visit the space station who was not a member of ESA at the time.[3] Like STS-100, STS-111 delivered part of the SSRMS and carried an MPLM, this time *Leonardo*. He conducted the first spacewalk by a European astronaut at the ISS. After he resigned from ESA, he became a test pilot for Airbus.

Belgian Frank de Winne flew on Soyuz TMA-1 in October 2002. The Odissea mission was his first spaceflight and he returned in the capsule that Roberto Vittori had delivered. He would go on to be ESA's first ISS commander on a later mission.

---

[1] Umberto Guidoni's STS-100 mission was not given a name by ESA, but each subsequent mission had a unique ESA name, carrying on from the tradition started with CNES and ESA missions to Mir.

[2] The Italian Space Agency (ASI) and the Italian aerospace firm, Thales Alenia Space, can rightly claim to have built a significant amount of the International Space Station. Their products include the Tranquility and Harmony module built for NASA, the Columbus module built for ESA, the Cupola for ESA and the Permanent Multi-Purpose Module *Leonardo*. They also provided the Multi-Purpose Logistics Modules, the pressurized cargo section of the Automated Transfer Vehicles and the pressurized cargo module of the Cygnus spacecraft. For this reason, ASI has bartered a number of spaceflights for it astronauts, out of proportion to its influence in ESA.

[3] As usual, there are exceptions that prove the rule. Charles Simonyi was born in Hungary and Richard Garriott was born in the UK. Both were US citizens and flew to the station as space tourists.

After the grounding of the space shuttles following the loss of Columbia on the STS-107 mission, the station was reduced to a skeleton crew of two and the Soyuz became the only means of getting to and from the ISS. Spaniard Pedro Duque flew to the station on Soyuz TMA-3 with Alexandr Kaleri of Roscosmos and Michael Foale of NASA. They remained on the station and he returned on Soyuz TMA-2 with Yuri Malenchenko and Ed Lu. The October 2003 Cervantes mission was Duque's second and last spaceflight. Like some of his ESA colleagues, he entered politics in his home country and is currently the Minister of Science in the Spanish government.

Dutch medical doctor, André Kuipers, flew to the station on Soyuz TMA-4 in April 2004 and returned with Kaleri and Foale. His mission was called DELTA and he would fly once more as part of a long duration crew.

Vittori returned to the station on his Eneide mission in April 2005 and he was followed by ESA's first long duration mission on the ISS. Already a veteran of a long duration mission on Mir, German pilot Thomas Reiter spent 171 in space as part of the Expedition 13 and 14 crews. His Astrolab mission was from July to December 2006. He launched on STS-121 and returned on board STS-116. As on his Mir mission, he conducted a spacewalk during his time at the station.

The next three ESA missions were space shuttle construction and supply missions with relatively short stays on board for the European astronauts. Swedish physicist Christer Fuglesang, on his Celsius mission, was part of the STS-116 crew that delivered part of the solar array truss to the station in December 2006. He participated in three of the mission's four spacewalks.

He was followed by Italian Paolo Nespoli on the Esperia mission on board STS-120. This mission expanded the station considerably when it delivered the Node 2 Harmony module in October 2007. Nespoli would visit the station twice more as a member of the long-term crew.

The February 2008 flight of Atlantis, STS-122, carried the hopes of ESA in its payload. On this mission, Hans Schlegel of Germany and Léopold Eyharts of France accompanied the ESA Columbus laboratory. This was the major contribution of ESA to the station. Schlegel performed a spacewalk during the mission, unrelated to the Columbus installation, and returned on STS-122 after 12 days in space. Eyharts remained on the ISS for a "mini-expedition" spending 48 days as part of Expedition 16 and returning on STS-123 in March 2008.

History was made again on the next ESA mission to the station, when Frank de Winne became the first European to command the facility. His Oasiss mission encompassed Expeditions 20 and 21. He commanded the latter, returning to Earth in December 2009. This was his second and last spaceflight and he returned to ESA to become the head of the European Astronaut Centre.

Christer Fuglesang returned to the station on his Alissé mission in August 2009. Travelling on STS-128, he conducted two more spacewalks to prepare the station for the arrival of the Tranquility module. After his final flight, he returned to

Sweden to take up the post of professor of Space Travel at KTH Royal Institute of Technology and director of the KTH Space Center.

Paolo Nespoli returned to the station in December 2010 on his Magisstra mission. This was his second spaceflight but his first long duration mission. He arrived on a Soyuz this time, Soyuz TMA-20, and stayed until May 2011 as a member of Expeditions 26 and 27.

Roberto Vittori returned to the station for a third time in May 2011, a first for an ESA astronaut. He was on board STS-134 as it delivered the most complex and most expensive scientific instrument ever launched to the ISS, the Alpha Magnetic Spectrometer (AMS-02).[4] He was also the last ESA astronaut to fly on the space shuttle as STS-134 was the programme's penultimate mission and the last flight for Endeavour. His mission was named DAMA.

André Kuipers returned to the station as a member of Expeditions 30 and 31 on his Promisse mission. He launched on Soyuz TMA-03M in December 2011 and returned in July 2012 after 192 days in space.

The next European astronaut to fly in space was the first member of the ESA 2009 astronaut class, Luca Parmitano, on his Volare mission. This is where this book takes up the story of European missions to the International Space Station. It covers the adventures of that new cohort of spacefarers: Luca, Alexander, Samantha, Andreas, Tim and Thomas and brings us up to date with Parmitano's return to the station to take command on his Beyond mission in 2019.

---

[4] AMS-01 was a prototype that flew on STS-91 in June 1998.

2

Spacecraft Flown to the ISS

European astronauts have traveled to the ISS in two types of spacecraft: the American space shuttle and the Russian Soyuz. They have been resupplied on board the station by payload in the Italian-built Multi-Purpose Logistics Modules (MPLM) carried aboard shuttles and also by a variety of uncrewed vehicles – the Russian Progress, the European ATV, the Japanese HTV, the SpaceX commercial Dragon spacecraft and Orbital Sciences Cygnus.[1]

Each type of flight to the ISS uses a different code:

R: Russian Roscosmos flight
A: USA NASA flight
E: European ESA flight
J: Japanese JAXA flight
A/R: Joint USA/Russian flight (financed by USA, built by Russia)
J/A: Joint Japanese/USA flight
UF: Utilization flight
LF: Logistics flight
ULF: Utilization/Logistics flight
S: Crew delivery flight on Soyuz
P: Cargo delivery flight on Progress
ATV: Cargo delivery flight on ESA Automated Transfer Vehicle
HTV: Cargo delivery flight on JAXA H-II Transfer Vehicle
SpX: Cargo delivery flight on SpaceX Dragon
Orb: Cargo delivery flight by Orbital Science Cygnus.

---

[1] Descriptions of the various spacecraft have been adapted from the author's first book, *In the Footsteps of Columbus, European Missions to the International Space Station,* which covers the period 2001–2012.

© Springer Nature Switzerland AG 2020
J. O'Sullivan, *European Missions to the International Space Station,*
Springer Praxis Books, https://doi.org/10.1007/978-3-030-30326-6_2

The Space Shuttle

During the period covered in this book (2013–2019), none of the European astronauts were transported to the station on the space shuttle. The Space Transportation System (STS) was retired after Atlantis touched down at the Shuttle Landing Facility at the Kennedy Space Center on 21 July 2011. The STS-135 mission had been the final resupply mission to the ISS, using the MPLM *Raffaello* to transport the payload. The previous mission, STS-134, carried ESA and ASI astronaut Roberto Vittori in May 2011.

Fig. 2.1. Space shuttle *Discovery,* STS-128 (NASA)

Soyuz

The Soyuz spacecraft has been operational in many guises since 1967 and is still the mainstay of human spaceflight to the ISS. It was developed by the Soviet Union as the successor to the Vostok/Voskhod spacecraft, and because it could carry three people it was roughly comparable to NASA's Apollo spacecraft. Although uncrewed variants (named Zond) made circumlunar flights, Soyuz was never used to fly a crew into lunar orbit. Nevertheless, it has been used for Earth orbital operations, the ASTP docking, and ferrying cosmonauts to and from the Salyut, Mir, and ISS space stations.

The Soyuz spacecraft is launched on the eponymous rocket from the Baikonur Cosmodrome in Kazakhstan and consists of three parts:

- The Orbital Module that provides accommodation for the crew during their mission.
- The Descent Module that carries the crew into orbit and returns them to Earth.
- The Service Module that contains the instruments and engines and has solar panels attached.

There have been many variants of the Soyuz spacecraft since its inception, including the first generation crewed Soyuz 7K-OK, the Soyuz 7K-OKS for ferry missions to Salyut, the second generation ferry Soyuz 7K-T and the Soyuz 7K-TM that docked with the American Apollo capsule during the 1975 Apollo-Soyuz Test Project (ASTP).

There followed continuous development from the third generation Soyuz T, the fourth generation Soyuz TM, which flew to the Mir space station and to the ISS, the Soyuz TMA which featured a 'glass' cockpit and whose couches could accommodate taller cosmonauts (notably Americans), the Soyuz TMA-M with digital control systems, and the final variant, known as Soyuz MS, that is in use today. Russia intends soon to retire Soyuz in favor of an entirely new spacecraft.

On ISS missions, European cosmonauts flew on the later Soyuz TM and TM models. For the period covered in this book the ESA spacefarers travelled on the TMA-M and MS models.

Fig. 2.2. The Soyuz TMA-7 'crew taxi' (Wikipedia)

Progress

Progress is an uncrewed version of Soyuz and shares its architecture and design. It is a cargo freighter and has been used to deliver supplies to the Salyut, Mir, and ISS space stations. The Descent Module of the Soyuz was replaced with an unpressurized propellant and refueling compartment. It can deliver up to 2230 kg of cargo to the ISS, to which it docks automatically.

Fig. 2.3. A Progress freighter approaching the ISS (NASA)

The Progress engines can be used to boost the ISS to orbit. The entire craft burns up on re-entering the atmosphere, taking away trash from the station.

ATV

The Automated Transfer Vehicles are ESA's uncrewed cargo spacecraft, five of which were launched to the ISS between 2008 and 2014. The design was based on the MPLM module, fitted with a propulsion system. It docked automatically with the Russian section of the ISS and could deliver up to 7667 kg of cargo.

Fig. 2.4. ATV-2, 'Johannes Kepler' (ESA)

Like Progress, the ATV's engines could be used to boost the ISS to orbit. It would burn up on re-entry, taking away trash from the station.

H-II

The H-II Transfer Vehicle is JAXA's uncrewed cargo spacecraft. The design consists of four parts:

- The Pressurized Logistics Carrier (PLC) that contains the Common Berthing Mechanism to mate with the ISS and enable station crew to gain access.
- The Unpressurized Logistics Carrier (UPLC) that contains the Exposed Pallet, which can be transferred to the exterior of the ISS by robotic arm.
- The Avionics Module.
- The Propulsion Module.

The HTV has a payload of 6000 kg, including 5200 kg carried in the PLC. An HTV doesn't dock automatically; it approaches the ISS and is grappled by the station's robot arm and berthed by the ISS crew. It burns up on re-entry, taking away trash from the station. Eight vehicles were launched between 2009 and 2019, and at the time of writing this book, there are plans for two more to be launched between 2020 and 2022.[2]

---

[2] HTV-8 was berthed on September 28, 2019, 2 days before this manuscript was submitted for editing.

Fig. 2.5. An HTV grappled by Canadarm2 (NASA)

Dragon

The Dragon spacecraft was developed by SpaceX as part of NASA's Commercial Resupply Services (CRS) programme where commercial companies design, build, and operate vehicles under contract.[3] This was a departure from the previous model where the government, via NASA, own the spacecraft.

Fig. 2.6. A Dragon capsule grappled by Canadarm2 (NASA)

The design consists of two parts:

- The blunt-cone pressurized ballistic capsule that can return to Earth and is re-usable.
- The unpressurized cargo-carrier trunk section that is equipped with two solar arrays.

The Dragon approaches the ISS and is grappled by the station's robot arm and berthed by the ISS crew. It can deliver up to 3310 kg of cargo. In a valuable service, it can also return cargo to Earth. A total of eight were launched to the ISS between 2010 and 2015; all successful apart from the last one, which was lost when the launch vehicle failed. Flights resumed in April 2016, and (at the time of writing)

---

[3] In 2008 SpaceX and Orbital Sciences were awarded contracts for twelve and eight resupply missions to the ISS, respectively. These were extended in 2015 to twenty and ten missions, respectively. In 2016, six CRS-2 contracts were awarded each to SpaceX, Orbital ATK and Sierra Nevada.

there have been an additional ten missions. With the second CRS contract in place, flights are planned to continue into 2024.[4]

Cygnus

After the failure of Rocketplane Kistler to meet its obligations under the Commercial Orbital Transportation Services (COTS) contract, NASA appointed Orbital Sciences to join SpaceX as the second commercial resupplier to the ISS using the Cygnus spacecraft.[5]

The Cygnus consists of two parts:

- The Pressurized Cargo Module (PCM), built by Thales Alenia in Italy. Like the Progress, it burns up on re-entry.
- The Service Module (SM) built on Orbital ATK GEOStar and LEOStar satellite buses. It is equipped with two solar arrays.

Fig. 2.7. A Cygnus capsule grappled by Canadarm2 (NASA)

---

[4]At the time of publication (Spring 2020) NASA had announced the first crews to fly to the ISS on two commercial vehicles. Robert Behnken and Douglas Hurley are to fly on a SpaceX Dragon 2 in May 2020. Christopher Ferguson, Michael Finke (replacing Eric Boe) and Nicole Mann are scheduled to fly on a Boeing CST-100 Starliner in Q2 2020.

[5]Orbital Sciences merged with Alliant Techsystems (ATK) in February 2015, and in June 2018 Northrop Grumman purchased Orbital ATK, which became a division of the parent company and renamed Northrop Grumman Innovation Systems.

The first four Cygnus spacecraft used the standard PCMs with a volume of 18 m^3. Later enhanced variants have had a volume of 27 m^3. The Enhanced model can deliver up to 3200 kg of cargo the ISS. In another parallel with SpaceX, a Cygnus was lost in a launch explosion in October 2014. However, flights resumed in December 2015 and will continue under the second CRS contract.

Part II

Missions

3

Volare

Mission
ESA Mission Name:	Volare
Astronaut:	Luca Parmitano
Mission Duration:	166 days, 6 hours, 17 minutes
Mission Sponsors:	ESA/ASI
ISS Milestones:	ISS 35S, 66th crewed mission to the ISS

Launch
Launch Date/Time:	May 28, 2013, 20:31 UTC
Launch Site:	Pad 1, Baikonur Cosmodrome, Kazakhstan
Launch Vehicle:	Soyuz TMA
Launch Mission:	Soyuz TMA-9M
Launch Vehicle Crew:	Fyodor Nikolayevich Yurichikhin (RKA), CDR
	Luca Salvo Parmitano (ESA), Flight Engineer
	Karen Lujean Nyberg (NASA) Flight Engineer

Docking
Soyuz TMA-9M
Docking Date/Time:	May 29, 2013, 02:10 UTC
Undocking Date/Time:	November 10, 2013, 23:26 UTC
Docking Port:	Rassvet nadir

Landing
Landing Date/Time:	November 11, 2013, 02:49 UTC
Landing Site:	Near Dzhezkazgan, Kazakhstan
Landing Vehicle:	Soyuz TMA
Landing Mission:	Soyuz TMA-9M
Landing Vehicle Crew:	Fyodor Nikolayevich Yurichikhin (RKA), CDR
	Luca Salvo Parmitano (ESA), Flight Engineer
	Karen Lujean Nyberg (NASA) Flight Engineer

© Springer Nature Switzerland AG 2020
J. O'Sullivan, *European Missions to the International Space Station*,
Springer Praxis Books, https://doi.org/10.1007/978-3-030-30326-6_3

ISS Expeditions

ISS Expedition:	Expedition 36
ISS Crew:	Pavel Vladimirovich Vinogradov (RKA), ISS-CDR
	Aleksandr Aleksandrovich Misurkin (RKA), ISS-Flight Engineer 1
	Christopher John Cassidy (NASA), ISS-Flight Engineer 2
	Fyodor Nikolayevich Yurchikhin (RKA), ISS-Flight Engineer 3
	Luca Parmitano (ESA), ISS-Flight Engineer 4
	Karen Lujean Nyberg (NASA), ISS-Flight Engineer 5
ISS Expedition:	Expedition 37
ISS Crew:	Fyodor Nikolayevich Yurchikhin (RKA), ISS-CDR
	Karen Lujean Nyberg (NASA), ISS-Flight Engineer 1
	Luca Salvo Parmitano (ESA), ISS-Flight Engineer 2
	Oleg Valeriyevich Kotov (RKA), ISS-Flight Engineer 3
	Sergei Nikolayevich Ryazansky (RKA), ISS-Flight Engineer 4
	Michael Scott Hopkins (NASA), ISS-Flight Engineer 5

The ISS Story So Far

The previous ESA mission to the station was André Kuipers' Promisse mission,[1] which was launched aboard Soyuz TMA-04M in May 2012, when he joined Expedition 31. During that mission, the first SpaceX Dragon capsule arrived at the ISS, the first commercial resupply mission.

In June 2012, Shenzhou 9 docked with Tiangong-1. This was the first Chinese space station and became the second station in earth orbit. On the mission was the first Chinese woman in space, Liu Yang, and the first Chinese taikonaut to return to space, Jing Haipeng.

Soyuz TMA-05M delivered the crew of Expedition 32 and 33 in July 2012, including Japanese astronaut Akihiko Hoshide.

In October 2012, Soyuz TMA-06M launched from Pad 31/Site 6 at Baikonur to deliver the crew of Expedition 33. The last launch from that pad was Soyuz T12 in July 1984. Pad 1/Site 5, known as 'Gagarin's Star', was being refurbished.

In December 2012, Soyuz TMA-07M delivered three crewmembers to the station to join Expedition 34. Canadian Chris Hadfield was making his third spaceflight and his second visit to the station. He joined Expedition 34 as a flight engineer, becoming the second Canadian member of an expedition. He commanded Expedition 35 in March 2013, becoming the first, and so far, only Canadian to command the station. His video of his cover recording of the David Bowie song *Space Oddity*, caused a social media sensation in that year.[2]

---

[1] See this author's book, *In the Footsteps of Columbus, European Missions to the International Space Station.*

[2] The video, *Space Oddity,* has over 44 million views on YouTube at the time of writing.

Soyuz TMA-08M docked at the ISS in March 2013 after the fastest rendezvous of a Soyuz up to that time. It lasted less than 6 hours.[3]

Luca Parmitano

Early Career

Luca Parmitano was born on September 27, 1976, in Paternò, Sicily. He graduated from the Liceo Scientifico Statale 'Galileo Galilei' in Catania, Italy, in 1995. As an exchange student, he studied at Mission Vejo High School, California, and went on to gain a degree in political sciences at the University of Naples Federico II, studying international law.

He joined the Italian Air Force and graduated from the Accademia Aeronautica, Pizzuoli, in 2000. His assignments began with Euro-NATO Joint Jet Pilot Training at Sheppard Air Force Base, Texas, in 2001. He flew AMX ground attack jet fighters with the 13° Gruppo, 32° Stromo in Amendola, Italy, from 2001 to 2007, earning qualifications, including Combat Ready, day/night air to air refueling, Four Ship Leader and Mission Commander/Package Leader. In 2002 he completed the JCO/CAS course with the USAFE in Sembach, Germany. In 2003, he qualified as Electronic Warfare Officer at the ReSTOGE in Pratica di Mare, Italy. He completed the Tactical Leadership Programme in Florennes, Belgium, in 2005. He went on to serve as Chief of Training, 76th flight commander and 32° Stromo (Wing) Electronic Warfare Officer (EWO).

In 2007 he qualified as a test pilot from the École du Personnel Navigant D'essais et de Reception (EPNR) test pilot school at Istres, France, and in July 2009 he was awarded a master's degree in experimental flight test engineering from the Institute Superieure de l'Aeronautique et de l'Espace (ISAE) in Toulouse, France. Parmitano is still a Lieutenant Colonel in the Italian Air Force. He has logged over 2000 hours flying time, is qualified on more than 20 types of military airplanes and helicopters, and has flown over 40 types of aircraft. He was selected as an ESA astronaut as part of the 2009 class.

---

[3] ESA astronaut Paulo Nespoli explained to the author at a talk in Cork, Ireland, in July 2018 that this was due to the updated processor, communications and satellite navigation equipment on the latest Soyuz variants. Communication is no longer reliant on an overhead radio line-of-sight; on board computers are more powerful, and GPS navigation allows more accurate location.

Fig. 3.1. Luca Parmitano (ESA)

Parmitano married Kathy Dillow, and has two daughters. He enjoys scuba diving, snowboarding, skydiving, weight training and swimming as well as reading and listening to water music.

He has received the following honors:

- Silver Medal to the Aeronautical Valour by the president of the Italian Republic, 2007.[4]
- Commendatore al Merito della Repubblica by the president of the Italian Republic, 2013.
- Asteroid 37627 Lucaparmitano is named after him.

---

[4] Awarded after landing his AMX fighter safely after a bird strike.

The Volare Mission

Mission Patches

The Volare mission patch was designed by Ilaria Sardella, 28, from San Giorgio Ionico, Italy. It shows the Soyuz spacecraft and the ISS orbiting Earth, with stylized orbit in the colors of the Italian flag. The orbits represent our desire to travel beyond Earth and the Sun, as well as our thirst for knowledge. Norberto Cioffi, 32, from Milan, chose the name Volare, meaning 'to fly', as part of a competition.

The Soyuz TMA-09M patch was designed by Luc van den Abeelen, Dmitri Shcherbinin and Fyodor Yurchikhin. It shares a common design with the patches of TMA-10 and TMA-19. Yurchikhin flew on both. In all three designs a light blue outer circle, containing the crewmember's names, surrounds a center circle including the Earth and the spacecraft in yellow and orange. The 'cog' shape represents the outer rim of the Vizir Spetsialniy Kosmicheskiy-4 (VSK-4) periscope on the Soyuz. The four white and pale blue stripes underneath the Soyuz symbolize the fourth mission to the station for Yurchikhin, and the station graphic is shared with the patch of Expedition 37, which he was to command.

Fig. 3.2. Volare mission patch (ESA)

Fig. 3.3. Soyuz TMA-09M mission patch (ESA)

Fig. 3.4. Expedition 36 mission patch (NASA)

Fig. 3.5. Expedition 37 mission patch (NASA)

The Expedition 36 patch was designed by Blake Dumesnil. The almost art deco design highlights the ISS's famous solar arrays. The NASA press release explained that "slanted angles denote a kinetic energy leading from the Earth in the lower right to the upper left tip of the triangular shape of the patch, representing the infinite scientific research, education, and long-duration spaceflight capabilities the ISS provides with each mission, as well as our goal for future exploration beyond the Space Station. The numbers 3 and 6 harmoniously intertwine to form expedition number 36 and its gray coloration signifies the unity and neutrality among all of the international partners of the ISS. The blue and gold color scheme of the patch represents the subtle way the central gold orbit wraps around the number 36 to form a trident at its lower right tip. The trident also symbolizes the sea, air, and land, all of which make up the Earth from where the trident originates in the design." It has also been suggested that it honors Chris Cassidy, a Navy SEAL, as there is a trident in the SEAL's insignia.

The Expedition 37 patch is dominated by Leonardo da Vinci's Vitruvian Man. this 525-year-old image combines art, science and medicine. This is apt, as Kotov is a physician and Ryazansky is a biochemist. Also, Parmitano, like da Vinci is Italian. The six stars represent the six members of Expedition 37 crew.

Fig. 3.6. The Soyuz TMA-09M crew with Luca Parmitano on right (www.spacefacts.de)

Fig. 3.7. The Expedition 36 crew with Luca Parmitano third from right (NASA)

Fig. 3.8. The Expedition 37 crew with Luca Parmitano third from left (NASA)

Timeline

Expedition 36, Week Ending June 2, 2013

Soyuz TMA-09M launched from Baikonur at 22:31 CET on May 28, 2013, carrying Fyodor Yurichikhin of Roscosmos, Luca Parmitano of ESA and NASA's Karen Nyberg. They reached orbit 9 minutes later and the spacecraft docked with the station's Rassvet port at 04:17 CET. Station commander Pavel Vinogradov, Alexander Misurkin and Chris Cassidy welcomed the trio to the station, and they joined Expedition 36.

This fast-track rendezvous procedure was enabled by recent upgrades in navigation and communication equipment, and Parmitano was the first ESA astronaut to take advantage. Parmitano was the first of the ESA class of 2009 to fly in space.[5]

---

[5] The six astronauts selected in 2009 were Samantha Cristoforetti of Italy, Alexander Gerst of Germany, Andreas Mogensen of Denmark, Luca Parmitano of Italy, Timothy Peake of United Kingdom and Thomas Pesquet of France. Matthias Maurer reached the final 10 in 2009, but was not selected and joined ESA as a crew support engineer. He joined the astronaut corps in July 2015 and completed basic training in 2018.

Cassidy shaved his head in an effort to welcome the similarly shaved Parmitano. "I don't think I've looked like this since Plebe Summer!" said Cassidy, referring to the 7-week induction to the U. S. Naval Academy.[6]

On May 30, Cassidy set up the Microgravity Science Glovebox to conduct the Burning and Suppression of Solids (BASS) experiment. This experiment examines how solid materials ignite in microgravity, and may lead to new approached to dealing with fires aboard spacecraft. He also deployed dosimeters throughout the station to measure radiation to which the crew may be exposed. He logged his meals for the Pro K experiment. This is to study how diet can reduce the bone loss experienced by astronauts in space. In the Kibō module, Nyberg conducted a leak test on the nitrogen gas supply line to the Multi-purpose Small Payload Rack's combustion chamber.

Parmitano set up part of the wireless network aboard the station and added an accelerometer to record data from the Russian treadmill. Vinogradov replaced connectors on pumps in the Russian segment and recorded data from the Kulonovskiy Kristall experiment, studying charged particles in a weightless environment. Misurkin carried out the Matryoshka experiment, which analyzes the radiation on the station. Misurkin also executed the Relaksatsiya Earth-observation experiment which examines chemical luminescent reactions in the Earth's upper-atmosphere. Yurchikhin transferred cargo out of the Soyuz and dried the Sokol launch and entry suits and gloves he and his crewmates wore on their way to the station.

The crew photographed Tropical Storm Barbara off the Pacific coast of Mexico.

On Friday, May 31, Nyberg and Parmitano measured how air circulates through the ISS. The Temperature and Humidity Control Intermodule Ventilation system prevents carbon dioxide from building up in microgravity. Nyberg then measured the insulation resistance of the Gradient Heating Furnace in Kibō's Kobairo rack. This is a vacuum chamber containing three heaters that generates high-quality crystals from melting materials. Cassidy performed maintenance on the Water Recovery System, changing the recycle tank. As normal, shortly after new crewmembers arrived, emergency procedures were reviewed.

Vinogradov loaded trash on to the ISS Progress 51 cargo ship docked to the aft docking port of the Zvezda module, before it was to depart and burn up on re-entry, making way for the ESA 'Albert Einstein' Automated Transfer Vehicle, ATV-4. Misurkin conducted another session of the Relaksatsiya Earth-observation experiment and Yurchikhin continued transferring cargo out of the Soyuz.

---

[6] Cassidy wore a fake moustache when he first arrived on the station to join Expedition 35 in March 2013. This was in honor of station commander Chris Hadfield, who is famous for his moustache.

Expedition 36, Week Ending June 9, 2013

The ESA Automated Transfer Vehicle (ATV-4) launched on an Ariane 5 rocket from Kourou, French Guiana, at 5:52 p.m. EDT on June 5. The 13-ton ATV carried 2478 kg of cargo, 860 kg of propellant for the Zvezda service module, 2580 kg of propellant for reboost and debris avoidance, 570 kg of water and 100 kg of oxygen and air.

Cassidy, Nyberg and Parmitano carried out eye examinations and downlinked the data for analysis by medical ground support teams to study the effect of microgravity on sight. Cassidy maintained the Multi-purpose Small Payload Rack Combustion Chamber and cleaned ventilation grills in Kibō. Nyberg and Parmitano received station orientation training from Cassidy. Vinogradov and Misurkin carried out tests as part of the Bar experiment, which attempts to detect air leaks on the station.

Expedition 36, Week Ending June 16, 2013

Progress 51 undocked from the aft port of the Zvezda module on June 11 at 9:58 a.m. EDT. Its cameras recorded the Zvezda's docking port sensors to establish if they were damaged when the Progress arrived. One of the KURS automated rendezvous antennas had not deployed after launch. The sensors were shown to be undamaged and ready for the arrival of the ATV. The ATV 'Albert Einstein' docked Saturday at 10:07 a.m. EDT to the aft-end port of the Zvezda service module.

Expedition 36, Week Ending June 23, 2013

On June 17, Vinogradov and Parmitano performed leak checks between the Zvezda service module and the Albert Einstein ATV-4. Mission control postponed the original hatch opening as some cargo was suspected of harbouring mold. Yurchikhin and Misurkin prepared for their June 24 spacewalk by checking the Russian Orlan spacesuits and studying the worksite and planned movements. Yurchikhin carried out the Seiner ocean-observation experiment and Misurkin prepared microbial test kits. Nyberg maintained the Tranquility and Destiny laboratories. She then joined Parmitano to discuss the Spinal Ultrasound experiment. This images the spine to investigate the fact that astronauts grow taller by up to 3% during long duration missions. Parmitano also conducted routine maintenance on the Advanced Resistive Exercise Device (ARED). Cassidy controlled the Surface Telerobotics K10 rover at the Ames Research Center in California from the ISS to simulate deploying a radio telescope. This was the first time an astronaut remotely controlled a robot on a planet's surface.

At 4:40 a.m. EDT, June 18, Vinogradov and Parmitano opened the hatch to the Albert Einstein and conducted a video survey. They used ventilation fans to clean the air to enable crewmembers to safely enter to unpack the 7.3 tons of supplies. Parmitano talked to Isabelle Kumar of Euronews and Cassidy carried out an ultrasound scan on Nyberg for the Spinal Ultrasound investigation. Afterwards, Nyberg tested samples of the Water Recovery System with the Total Organic Carbon Analyzer to check for microbial contaminants. Cassidy inspected the fire extinguishers and portable breathing apparatuses, and recharged the batteries of the American EVA suits. He also adjusted the video camera and lighting rig for the Capillary Channel Flow experiment inside the Microgravity Science Glovebox to improve the video transmissions for ground controllers. Misurkin and Yurchikhin resized their Russian Orlan EVA suits checked the suits' valves.

The crew continued to photograph Earth as part of the Crew Earth Observations program. They captured images of Kingston, Jamaica; Ulaanbaatar, Mongolia; and Moroni, Comoros.

The ISS Progress 51 cargo craft had earlier undocked from the station on June 11. Its place was taken by the Albert Einstein ATV. Russian ground controllers carried out a week of engineering tests on the Progress before it re-entered the atmosphere.

Cassidy, Nyberg and Parmitano unloaded cargo from the ATV on June 19, including the Fundamental and Applied Studies of Emulsion Stability (FASES). This was installed in the Fluid Science Laboratory to investigate the behavior of emulsions, a combination of two liquids that do not mix well, in microgravity. Cassidy installed a new rope on the Advanced Resistive Exercise Device (ARED) and spoke to students at the Kansas Cosmosphere in Hutchinson, Kansas with Nyberg. Vinogradov installed a new docking mechanism in Pirs in preparation for the departure of Progress 50 and the arrival of Progress 52. He then maintained on the toilet in the Russian segment. He downloaded data from the Identification experiment. This measures dynamic stress during dockings and reboosts. Misurkin and Yurchikhin studied airlock procedures for their upcoming EVA and charged batteries on their Orlan suits.

At 9:05 a.m. the ATV-4's thrusters fired for 6 minutes, 47 s. This increased the station's altitude by 0.3 miles at apogee and 1.9 miles at perigee and left the ISS at an altitude of 265.5 × 251.0 miles. At almost the same time, Progress 51 was deorbited over the Pacific Ocean. It had been conducting freeflight tests and acting as a target for ground-based radar for the previous week.

On June 20, Nyberg checked in on the Constrained Vapor Bubble experiment inside the Fluids Integrated Rack, inspecting and photographing the scientific payload. This experiment takes a look at the physics of evaporation and condensation in the absence of gravity. Parmitano meanwhile reached the midpoint of his participation in the Biological Rhythms 48hrs experiment, a Japan Aerospace Exploration Agency (JAXA) study of the circadian variation of heartbeat during spaceflight using a digital electrocardiograph. He ended a 24 hours session and started a new one. On this day, Nyberg and Parmitano also logged their meals for the Pro K study, which postulates that a diet with a decreased ratio of animal protein to potassium results in lower mineral bone-loss during long duration space missions. Cassidy took an inventory of the batteries and small propellant tanks of the three Synchronized Position Hold, Engage, Reorient, Experimental Satellites (SPHERES). These are free flying soccer ball-sized satellites that fly inside the station. Cassidy, Nyberg and Parmitano finished the day unloading cargo from the ATV. Yurchikhin, Misurkin and Vinogradov closed the hatch between Progress 50 and the Pirs docking compartment and moved the Orlan spacesuits into Pirs to prepare for a practice suit up.

Nyberg and Parmitano took blood samples on June 21 for the Human Research Facility. This measures adaptation to microgravity. Parmitano processed the blood samples and stored them in the Minus Eighty-degree Laboratory Freezer for ISS, or MELFI. Nyberg worked on the DEvice for the study of Critical LIquids and Crystallization (DELIC). This examines the phase transitions of transparent substances in microgravity. Nyberg installed the High Temperature Insert to examine water as it crystallizes. DELIC aids the design of supercritical water reactors to treat waste which could lead to new innovations in energy and waste treatment. She worked on the Optical Coherence Tomography equipment which records hi-res images of the eye. Parmitano completed his session with the JAXA Biological Rhythms 48hrs. Parmitano downloaded data from his medical monitors and from those worn by Nyberg during the previous session. Cassidy prepared for an experiment using the Binary Colloid Alloy Test-C1 experiment. Colloids are microscopic particles suspended in liquid. Colloid research could lead to increased longevity of food and medicine. He also installed a camera in the Quest airlock in advance of his spacewalk with Parmitano, scheduled for later in July.

Expedition 36, Week Ending June 30, 2013

Yurchikhin and Misurkin began a 6 hours, 34 minutes spacewalk at 9:32 a.m. June 24. The EVA was to prepare for the addition of the 'Nauka' Multipurpose Laboratory Module, which was planned to replace Pirs.[7] They replaced a fluid flow control panel on the Functional Cargo Block, FGB. This was ongoing preventative maintenance on the cooling system for the Russian segment of the station. They also attached power cable brackets for the Nauka and replaced external science experiments. This was the 169th spacewalk in support of space station

---

[7] In 2013, Nauka was scheduled to be launched later the same year. However, at the time of writing (September 2019), it had not yet been delivered to the ISS. Rassvet was the last Russian segment launched to the ISS. There are reports that Multi-purpose Laboratory Module (MLM) Nauka will be launched in late 2020, as contaminated tanks are being replaced. However, its completion has been significantly delayed because its originally planned launch date was 2007. If ever launched, it will replace Pirs at Zvezda's nadir port.

assembly and maintenance. It was Yurchikhin's sixth EVA and Misurkin's first one.

During the EVA, Cassidy and Vinogradov stayed in the Soyuz TMA-08M spacecraft docked at the Poisk module. Parmitano and Nyberg stayed in the U.S. segment. The Soyuz TMA-09M containing their couches, was docked to the Rassvet module on the Earth-facing side of the Zarya module. They carried out vision tests as part of the crew Health Maintenance System. Nyberg continued working with the Advanced Colloids Experiment.

On June 28, Parmitano worked with the humanoid Robonaut, and then he and Cassidy disassembled it. They downlinked the collected data to the controllers at the Marshall Space Flight Center. Cassidy removed and replaced the Potable Water Dispenser filter and tested it for leaks. Nyberg continued the unpacking of the ATV and conducted interviews with CBS *This Morning* and CNN's *Leading Women*. Vinogradov, Yurchikhin and Misurkin had a day off.

Expedition 36, Week Ending July 7, 2013

On July 1, Cassidy and Parmitano continued to prepare for their upcoming space-walks. To this end, they conducted training on the spacesuits Simplified Aid for EVA Rescue (SAFER) units and the Enhanced Caution and Warning System. In the Kibō module, Nyberg worked on the NanoRacks experiment. This is a commercial science platform which is controlled by students and businesses performing research. She also recorded ultrasonic background noise and replaced cables on a Multi-Purpose Small Payload Rack. Vinogradov, Misurkin and Yurchikhin photographed Earth's surface as part of a project to record the effects of industrial activity. They carried out Earth observations to attempt to predict potential manmade and natural disasters. Vinogradov performed maintenance on the Russian segment. He inspected the ventilation system of the Zarya cargo module and tested the communications systems. Misurkin unloaded cargo and updated the inventory. He helped Vinogradov with the communications tests. Yurchikhin inspected the Russian Orlan EVA suits.

In preparation for the upcoming EVAs, on July 2, Cassidy and Luca Parmitano checked out the glove heaters and rechargeable batteries of the U. S. spacesuits. The cosmonauts worked on plasma physics experiments. Vinogradov set up and photographed the Kulonovskiy Kristall experiment stowed outside the Zvezda service module. That study observes the plasma environment outside the space station. Yurchikhin worked with the Coulomb Crystal experiment, which studies electrically charged particles trapped in a magnetic field. Nyberg and Parmitano continued to unload cargo from the Automated Transfer Vehicle.

Spacewalkers Cassidy and Parmitano were helped by Nyberg, on July 3, with spacesuit fit checks in the U. S. Quest airlock. Later all three astronauts called down to the ground for a conference with spacewalk specialists. Vinogradov participated in an ongoing cardiac experiment with Misurkin and Yurchikhin.

Vinogradov also cleaned vents and replaced fan screens and airflow sensors and inspected and photographed Russian windows for analysis by specialists on the ground. Misurkin worked on the Molniya-Gamma experiment, which studies optical emissions in Earth's atmosphere and ionosphere associated with thunderstorm and seismic activity. He later conducted maintenance on the Elektron oxygen generation system. Yurchikhin installed two hoses in the Russian segment and checked laptop computers and updated software.

Expedition 36, Week Ending July 14, 2013

As Luca Parmitano prepared for his upcoming spacewalk, he told ESA controllers "This is one of the prime objectives of the mission, and it's really an honor to be able to do this, Siamo pronti ragazzi!" or "Guys, we're ready!" This would be the first spacewalk by an Italian astronaut and Parmitano had been coached by ESA veteran astronaut Christer Fuglesang, whose five EVAs are still a European record.[8]

Cassidy, Parmitano and Nyberg spent the morning of July 8 configuring spacewalk tools and equipment. Yurchikhin joined them for a final detailed procedure review. Afterward, the crew participated in a conference call with spacewalk specialists at Houston's Mission Control Center to address any remaining issues or questions. Nyberg set up the EarthKAM camera in the Window Observational Research Facility in the Destiny module. Students operate this digital camera to photograph the Earth's surface. Meanwhile, Vinogradov transferred cargo and waste to and from the Progress 50 resupply ship and later photographed the windows of the Zvezda service module to provide an assessment of their condition. Misurkin began working on the Khromatomass experiment, analyzing blood and saliva samples for this Russian health evaluation. He also prepared some radiation dosimeters for the spacewalk and participated in the Relaksatsiya Earth-observation experiment. This records chemical luminescent reactions in the atmosphere. Later, Yurchikhin updated software on a laptop computer and worked with the Kulonovskiy Kristall experiment. This examines charged particles in microgravity. The results of this study could be used on future spacecraft and advanced photovoltaic cells.

Cassidy and Parmitano began their spacewalk at 8:02 a.m. on July 9. Cassidy started on the Z1 truss to remove and replace a Space-to-Ground Transmitter Receiver Controller, which had not operated since December 2012. Parmitano, at the Express Logistics Carrier-2 on the starboard truss segment, retrieved two experiments from the Materials International Space Station Experiment-8 (MISSE-8). The Optical Reflector Materials Experiment III (ORMatE-III) and the Payload Experiment Container assessed the impacts of the space environment on

---

[8] Fuglesang performed two spacewalks during the Celsius mission on STS-116 and three during the Alissé mission on STS-128. All spacewalks were for ISS construction.

materials and processor elements. They were to return to Earth in a SpaceX Dragon capsule later in the year. Parmitano took pictures of the Alpha Magnetic Spectrometer-02 (AMS-02). These images would help ground controllers assess the state of the particle detector. Cassidy installed new cables to provide power to the anticipated Russian Multipurpose Laboratory Module, Nauka, that was scheduled to be launched later that year. Cassidy installed the cables from the Unity module to the interface between the Pressurized Mating Adapter-1 and the Zarya module. Together they removed two Radiator Grapple Bars (RGBs) and installed one on the port side truss and one on the starboard side.

Parmitano removed the Mobile Base Camera Light Pan-Tilt Assembly, which had failed the year before. Cassidy installed two Z1 truss Y-bypass jumpers to provide power redundancy and stability for critical station components. Parmitano installed a multi-layer insulation cover over the docking interface of Pressurized Mating Adapter-2 at the Harmony module.

As they were ahead of schedule, they were able to carry out some get-ahead tasks. They installed cables from the Zarya module to Pressurized Mating Adapter-1. Once back in the Quest airlock the EVA ended at 2:09 p.m. EDT, completing a 6 hours, 7 minutes spacewalk, Cassidy's fifth and Parmitano's first. This was the 170th spacewalk in support of station assembly and maintenance bringing the total to 1073 hours, 50 minutes.

The following day, July 10, was spent on post-spacewalk tasks. Cassidy and Parmitano undertook health exams conducted by crew medical officer, Nyberg. Then Cassidy recharged the water tanks of the U.S. spacesuits with iodinated water. Cassidy and Parmitano inspected and stowed tools used during the spacewalk and configured equipment and suits for the second EVA. Nyberg opened the shutter of the Window Observational Research Facility (WORF) in the Destiny laboratory for the ISS SERVIR Environmental Research and Visualization System (ISERV) experiment. This provides automated data acquisition and photographs for disaster monitoring. Misurkin retrieved a trio of Pille radiation dosimeters that had been placed inside the two U. S. spacesuits and inside a spacewalk equipment bag to record radiation exposure data.[9] Misurkin later collected a sample from the Vynoslivost experiment, which studies material fatigue in space. This provides data for estimating the lifespan of structural elements of the station. Yurchikhin worked on the Kulonovskiy Kristall experiment, investigating charged particles in microgravity and then helped Vinogradov examine the exterior surfaces of the Russian segment.

ATV-4 Albert Einstein was called on once more to raise the station's orbit. Thrusters were fired at 1:35 a.m. EDT for 9 minutes, 52 seconds to raise the station's orbit by 3.2 miles at perigee. This prepared the station for the arrival of Progress 52, due later in July and for the departure of the Expedition 36 crew on

---

[9] Russians have been using the Pille dosimeters since the Salyut 6 space station, which was in orbit between 1977 and 1982.

Soyuz TMA-08M, planned for September. The reboost left the station in an orbit of 263.7 × 254.7 statute miles with an average altitude of 259 statute miles.

July 11 saw Nyberg and Parmitano undergo standard health evaluations, such as the Reaction experiment. This examines how tiredness affects crew performance. Nyberg then had a fitness evaluation after the daily planning conference. She and Parmitano used the Cycle Ergometer with Vibration Isolation and Stabilization (CEVIS). Cassidy maintained the Binary Colloid Alloy Test experiment He replaced a filter and urine receptacle in the Waste and Hygiene Compartment. Parmitano and Nyberg replenished the Water Processor Assembly tanks. Vinogradov, Misurkin and Yurchikhin, answered questions from International Youth Science School students visiting the Russian Mission Control Center in Korolev, Moscow. Afterward, Vinogradov continued to stow trash aboard the Progress 50, ready for undocking on July 21.

At 7:37 a.m. the fire alarm in the Poisk module alerted the crew, but it was a false alarm.

After lunch, Cassidy, Parmitano and Nyberg reviewed the detailed timeline for the upcoming spacewalk in conjunction with Houston Mission Control Center. Nyberg replaced the battery of the EarthKAM camera in the Window Observational Research Facility in the Destiny laboratory. Misurkin installed samples in the Vynoslivost experiment. Yurchikhin returned to the Kulonovskiy Kristall experiment and also downloaded data from the Identification experiment. The crew also photographed Earth for the Crew Earth Observations program, including Typhoon Soulik in the Pacific Ocean, forest fires in Canada and the USA and floods in Europe.

On July 12, Parmitano spoke with Enrico Saggese, president of the Italian Space Agency (ASI) and reporters at the ASI HQ in Rome. He described his impressions of his first spacewalk and discussed the plans for the next one. Nyberg inspected the cooling systems in the Harmony node and the Destiny laboratory. Cassidy collected samples from the Tranquility node. Nyberg also prepared test samples for the Advanced Colloids Experiment (ACE). Vinogradov maintained the ventilation system in the Zvezda module and then collected tools and equipment for the Treadmill with Vibration Isolation System (TVIS). The TVIS was scheduled to be returned to the U. S. segment. Misurkin operated the Matryoshka experiment, which analyzes the radiation on the station. Later, he replaced panels in the Poisk Mini Research Module. Yurchikhin continued working on the Kulonovskiy Kristall experiment.

Expedition 36, Week Ending July 21, 2013

On July 16 Cassidy and Parmitano started their second spacewalk at 7:57 a.m. EDT. It was intended that they would continue preparing for the arrival of the Russian Nauka laboratory module as well as replace a video camera on the

Japanese Exposed Facility (JEF), move other cameras, investigate a troublesome door cover on the truss relay boxes and reinstall a thermal insulation cover on the truss.

Just over 1 hour into the EVA, Parmitano reported that water was floating inside his helmet. Ground controllers aborted the EVA immediately and Cassidy and Parmitano were back in the Quest airlock with the spacewalk complete at 9:39 a.m. EDT, or 1 hours, 32 minutes after starting. Luca described the experience in his ESA blog:

> *My eyes are closed as I listen to Chris counting down the atmospheric pressure inside the airlock – it's close to zero now. But I'm not tired – quite the reverse! I feel fully charged, as if electricity and not blood were running through my veins. I just want to make sure I experience and remember everything. I'm mentally preparing myself to open the door because I will be the first to exit the Station this time round. Maybe it's just as well that it's night time: at least there won't be anything to distract me.*
>
> *When I read 0.5 psi, it's time to turn the handle and pull up the hatch. It is pitch black outside, not the colour black but rather a complete absence of light. I drink in the sight as I lean out to attach our safety cables. I feel completely at ease as I twist my body to let Chris go by. In a matter of seconds, we finish checking each other and we separate. Even though we are both heading to more or less the same part of the International Space Station, our routes are completely different, set out by the choreography we have studied meticulously. My route is direct, towards the back of the Station, while Chris has to go towards the front first in order to wind his cable around Z1, the central truss structure above Node 1. At that moment, none of us in orbit or on Earth could have imagined just how much this decision would influence the events of the day.*
>
> *I pay careful attention to every move as I make my way towards the protective bag that we left outside the week before. I don't want to make the mistake of feeling so much at ease as to be relaxed. Inside the bag I find the cables that form part of what will perhaps be my most difficult task of the day. I have to connect them to the Station's external sockets while at the same time securing them to the surface of the station with small metal wires. Both operations involve me using my fingers a lot, and I know from experience that this will be really tiring because of the pressurised gloves.*
>
> *Chris partially connected the first cable last week, so I get hold of the part that is still unattached and I guide it carefully towards the socket. After a little initial difficulty, I inform Houston that I have completed the task and I'm ready for the second cable. After getting hold of the next cable, I move into what I think is the most difficult position to work from on the whole Station: I'm literally wedged between three different modules, with my visor and my PLSS (my 'backpack') just a few centimetres from the external walls of Node 3, Node 1*

and the Lab. Very patiently, with considerable effort I manage to fasten one end of the second cable to the socket. Then, moving blindly backwards, I free myself from the awkward position I've had to work in. On the ground, Shane tells me that I'm almost 40 minutes ahead of schedule, and Chris is also running ahead on his tasks.

At this exact moment, just as I'm thinking about how to uncoil the cable neatly (it is moving around like a thing possessed in the weightlessness), I 'feel' that something is wrong. The unexpected sensation of water at the back of my neck surprises me – and I'm in a place where I'd rather not be surprised. I move my head from side to side, confirming my first impression, and with superhuman effort I force myself to inform Houston of what I can feel, knowing that it could signal the end of this EVA. On the ground, Shane confirms they have received my message and he asks me to await instructions. Chris, who has just finished, is still nearby and he moves towards me to see if he can see anything and identify the source of the water in my helmet.

At first, we're both convinced that it must be drinking water from my flask that has leaked out through the straw, or else it's sweat. But I think the liquid is too cold to be sweat, and more importantly, I can feel it increasing. I can't see any liquid coming out of the drinking water valve either. When I inform Chris and Shane of this, we immediately receive the order to 'terminate' the sortie. The other possibility, to 'abort,' is used for more serious problems. I'm instructed to go back to the airlock. Together we decide that Chris should secure all the elements that are outside before he retraces his steps to the airlock, i.e., he will first move to the front of the Station. And so we separate.

As I move back along my route towards the airlock, I become more and more certain that the water is increasing. I feel it covering the sponge on my earphones and I wonder whether I'll lose audio contact. The water has also almost completely covered the front of my visor, sticking to it and obscuring my vision. I realise that to get over one of the antennae on my route I will have to move my body into a vertical position, also in order for my safety cable to rewind normally. At that moment, as I turn 'upside-down,' two things happen: the Sun sets, and my ability to see – already compromised by the water – completely vanishes, making my eyes useless; but worse than that, the water covers my nose – a really awful sensation that I make worse by my vain attempts to move the water by shaking my head. By now, the upper part of the helmet is full of water and I can't even be sure that the next time I breathe I will fill my lungs with air and not liquid. To make matters worse, I realise that I can't even understand which direction I should head in to get back to the airlock. I can't see more than a few centimetres in front of me, not even enough to make out the handles we use to move around the Station.

I try to contact Chris and Shane: I listen as they talk to each other, but their voices are very faint now: I can hardly hear them, and they can't hear me. I'm alone. I frantically think of a plan. It's vital that I get inside as quickly as possible. I know that if I stay where I am, Chris will come and get me, but how much time do I have? It's impossible to know. Then I remember my safety cable. Its cable recoil mechanism has a force of around 3lb that will 'pull' me towards the left. It's not much, but it's the best idea I have: to follow the cable to the airlock. I force myself to stay calm and, patiently locating the handles by touch, I start to move, all the while thinking about how to eliminate the water if it were to reach my mouth. The only idea I can think of is to open the safety valve by my left ear: if I create controlled depressurisation, I should manage to let out some of the water, at least until it freezes through sublimation, which would stop the flow. But making a 'hole' in my spacesuit really would be a last resort.

I move for what seems like an eternity (but I know it's just a few minutes). Finally, with a huge sense of relief, I peer through the curtain of water before my eyes and make out the thermal cover of the airlock: just a little further, and I'll be safe. One of the last instructions I received was to go back inside immediately, without waiting for Chris. According to protocol, I should have entered the airlock last, because I was first to leave. But neither Chris nor I have any problem in changing the order in which we re-enter. Moving with my eyes closed, I manage to get inside and position myself to wait for Chris' return. I sense movement behind me; Chris enters the airlock and judging from the vibrations, I know that he's closing the hatch. At that moment, communication passes to Karen and for some reason, I'm able to hear her fairly well. But I realise that she can't hear me because she repeats my instructions even though I've already replied. I follow Karen's instructions as best I can, but when repressurization begins I lose all audio. The water is now inside my ears and I'm completely cut off.

I try to move as little as possible to avoid moving the water inside my helmet. I keep giving information on my health, saying that I'm ok and that repressurization can continue. Now that we are repressurizing, I know that if the water does overwhelm me I can always open the helmet. I'll probably lose consciousness, but in any case that would be better than drowning inside the helmet. At one point, Chris squeezes my glove with his and I give him the universal 'ok' sign with mine. The last time he heard me.

The minutes of repressurization crawl by and finally, with an unexpected wave of relief, I see the internal door open and the whole team assembled there ready to help. They pull me out and as quickly as possible, Karen unfastens my helmet and carefully lifts it over my head. Fyodor and Pavel immediately pass me a towel and I thank them without hearing their words because my ears and nose will still be full of water for a few minutes more.

Space is a harsh, inhospitable frontier and we are explorers, not colonizers. The skills of our engineers and the technology surrounding us make things appear simple when they are not, and perhaps we forget this sometimes. Better not to forget.

Once safely inside the station the Post-Spacewalk briefing took place, including Flight Director, David Korth, lead spacewalk officer, Karina Eversley and ISS Mission Management Team Chairman, Kenneth Todd.

This was the second shortest EVA performed at the ISS.[10] It was the 171st spacewalk in support of station assembly and maintenance bringing the total to 1075 hours, 22 minutes.

Fig. 3.9. Luca Parmitano during the second EVA shortly before the water leak was discovered (ESA)

On Friday, July 19, Cassidy and Parmitano scrubbed of the cooling loops in the U. S. spacesuits. This maintenance task was planned before the aborted spacewalk. The two astronauts worked on Cassidy's suit and the spare but not on the suit worn by Parmitano. That suit was not cleared for EVA until it was discovered

---

[10] The shortest was the EVA of June 24, 2004, when Gennady Padalka and Michael Finke cut short their spacewalk after a pressure issue with Finke's primary oxygen tank. It lasted 13 minutes.

what had caused water to enter the helmet. Later, Parmitano spoke with the Italian Prime Minister Enrico Letta and discussed the events of the spacewalk. Cassidy, Vinogradov and Misurkin executed an emergency decent drill, practice for their upcoming departure. Afterwards, Cassidy reinstalled cables in the Destiny laboratory to provide power to the Robonaut, and cleaned smoke detectors in the Kibō laboratory. Nyberg maintained the Oxygen Generation System. She replaced a hydrogen sensor and cleaned an inlet for the Sabatier water extraction system. Vinogradov stowed more trash onto Progress 50 before its departure. Misurkin collected data from the Matryoshka experiment, while Yurchikhin carried out an inventory of the Zarya module and cleaned the Rassvet and Poisk ventilation systems.

Expedition 36, Week Ending July 28, 2013

The Progress 50 cargo ship undocked from the Pirs docking compartment on July 25 at 4:43 p.m. EDT, to be burned up on re-entry over the Pacific. Station commander Vinogradov and Yurchikhin practiced using the Telerobotically Operated Rendezvous System (TORU), a manual backup for use with the next arriving Progress in the event of a failure of the automatic system. Also, Nyberg prepared for another arriving cargo craft, the JAXA HTV-4, which does not dock automatically. It is captured and berthed using the Canadarm2 robotic arm. Cassidy set up the hardware for the Marangoni Inside experiment in the Fluids Physics Experiment Facility in the Kibō laboratory. He also collected urine samples. Parmitano replaced parts of the Combustion Integrated Rack. This records how different fuels burn in space. He installed neutron radiation detectors in the RaDI-N experiment. Misurkin worked on RaDI-N's predecessor, the Matryoshka radiation detection experiment.

Progress 52 launched from Baikonur, Kazakhstan at 4:45 p.m. EDT as the station was flying 260 miles over southern Russia, near the Kazakhstan and Mongolia border. Four orbits and 6 hours later, it docked at the station's Pirs docking compartment at 10:26 p.m. EDT, delivering 550 kg of propellant, 19 kg of oxygen, 28 kg of air, 420 kg of water, 1539 kg of spares and experiment hardware, including tools to repair the U. S. spacesuits.

Expedition 36, Week Ending August 4, 2013

The fourth JAXA H-II Transfer Vehicle (HTV-4), launched aboard an H-IIB launch vehicle from the Tanegashima Space Center, Japan at 3:48 p.m. EDT on 3 August, carrying more than 3.5 tons of cargo. This included research samples, a new freezer capable of preserving materials at temperatures below –90 F, four cubesats, food, water and supplies. The pressurized section also delivered new hardware for the Robotic Refuelling Mission.

Two Orbital Replacement Units (ORU), a spare Main Bus Switching Unit (MBSU), a spare Utility Transfer Assembly (UTA) and the Space Test Program – Houston 4 (STP-H4) payload were contained in the unpressurized section. The STP-H4 is a suite of experiments for investigating space communications, Earth monitoring and materials science.[11]

Expedition 36, Week Ending August 11, 2013

Parmitano, Nyberg and Cassidy used the robotics workstation in the station's cupola on August 5 to practice before the arrival of the HTV-4. They practiced the procedure to grapple and capture the craft with the Canadarm2. Once the dry run was complete, Mission Control remotely maneuvered the arm to a "high hover" position in readiness. Cassidy worked with Robonaut 2 in the Destiny laboratory. The team at NASA's Marshall Space Flight Center in Huntsville, Alabama, controlled the robot as it used a disinfectant wipe. Parmitano replaced cables on a camera port in the Quest airlock and later spoke with Italian press at the European Space Research Institute, Rome, Italy. In the Russian segment Yurchikhin and Misurkin prepared for their upcoming pair of spacewalks, scheduled for August, to install experiments and connect cables for the planned Russian Nauka Multipurpose Laboratory Module. They prepared their Russian Orlan spacesuits and configured systems inside the Pirs docking compartment. Vinogradov maintained the ventilation system of the Zvezda service module and audited medical kits.

Cassidy worked again on the Robonaut 2 on August 6. Wearing tele-operation vest, gloves and visor, he manipulated the Robonaut telerobotically by moving his own limbs, head and fingers. He then dismantled the robot. He also tested samples from the Water Processor Assembly (WPA) using the Total Organic Carbon Analyzer. This ensure uncontaminated water for the crew. Parmitano replaced a Pressure Control and Pump Assembly for the Urine Processing Assembly that had failed the previous week. Nyberg worked on the Marangoni Inside experiment in the Fluid Physics Experiment Facility of the Kibō laboratory. Nyberg and Cassidy spoke with NASA Administrator Charles Bolden, the press and public at NASA headquarters in Washington, D. C., to help celebrate the first anniversary of the Curiosity rover's landing on Mars. Nyberg also organised items stowed in the *Leonardo* Permanent Multipurpose Module in anticipation of the arrival of the HTV-4.

Nyberg continued to work on the Advanced Colloids Experiment (ACE) in the Destiny laboratory on August 7. She cleaned equipment and replaced a used oil dispenser. Afterwards, she joined Cassidy in the Kibō laboratory to prepare two SPHERES in preparation for the upcoming SPHERES Zero Robotics tournament scheduled for Tuesday, August 13. Parmitano set up acoustic dosimeters for

---

[11] Its predecessor, STP-H3, was delivered to the station during the final flight of space shuttle *Endeavour* on STS-134 in May 2011.

Vinogradov, Misurkin and Cassidy to wear for 24 hours to measure their exposure to noise. Then he gathered equipment to be installed in the vestibule of the Earth-facing port of the Harmony node, where the HTV-4 will be berthed. He also performed an ocular health assessment of Cassidy's eyes using an ultrasound scanner. Misurkin and Yurchikhin continued preparing their Russian Orlan spacesuits and spacewalking tools for their upcoming August spacewalks. Vinogradov maintained the Elektron oxygen generator and the Zvezda ventilation system.

On August 8, Nyberg set up the Capillary Flow Experiment to study interior corner fluid flow. She video recorded the results. Cassidy and Parmitano used the Cycle Ergometer with Vibration Isolation and Stabilization (CEVIS) as part of a routine fitness test which included blood pressure measurements. Later, Cassidy, Nyberg and Parmitano discussed the plans for the capture of the HTV-4 with flight controllers at the Johnson Space Center's Mission Control Center, Houston. Nyberg and Cassidy simulated using the Canadarm2 to capture the HTV-4. Nyberg then installed a new hardware command panel to monitor and control the approaching HTV-4. Cassidy removed a Space Acceleration Measurement System (SAMS) sensor from the Columbus laboratory's Microgravity Science Glovebox. Misurkin and Vinogradov repaired the Chibis-M suit, which simulates the gravity by reducing the pressure on the lower half of the body. Misurkin also installed samples in the Vynoslivost experiment.

Nyberg used the Canadarm2 to grapple the HTV-4 at 7:22 a.m. on August 9, as the uncrewed cargo ship came within about 30 feet of the station. Parmitano monitored events from the cupola. At the time of capture, the station was orbiting 260 miles south of South Africa. She controlled the robot arm to position the Japanese cargo ship in a ready-to-latch position on the Earth-facing port of the Harmony node. Nyberg and Cassidy initiated the bolting and first stage capture of Harmony's Active Common Berthing Mechanism (ACBM) with HTV-4's Passive Common Berthing Mechanism (PCBM). Then the mission controllers completed the bolting process through second stage capture. The HTV-4 delivered 3.6 tons of supplies.

Expedition 36, Week Ending August 18, 2013

On August 12, Nyberg, Cassidy and Parmitano had a day off but did have a briefing with ground robotic controllers to discuss the capture of the HTV-4. Vinogradov replaced batteries and mating telemetry connectors in the Zvezda service module. He conducted a hearing test, updated the inventory list and charged a battery for the Relaxation experiment. The latter examines the interaction of spacecraft and jet engine exhaust in the Earth's upper atmosphere and the station's environment. Yurchikhin and Misurkin inspected their Orlan spacesuits and checked for pressure leaks.

On August 13, Yurchikhin and Misurkin installed tools, lights and cameras to their Russian Orlan spacesuits. They reviewed procedures and tested the telemetry systems by communicating with flight controllers at the Russian Mission Control Center in Korolev, Moscow. Nyberg and Cassidy set up the SPHERES experiment in the Kibō laboratory for middle school students we were at the Massachusetts Institute of Technology for the micro-satellites simulation competition. Parmitano replaced a pre-treat tank on the Waste and Hygiene Compartment. Vinogradov charged batteries for a photo spectral system and carried out work on the Relaxation experiment. Misurkin prepared the bioreactor for manual mixing for the Cascade experiment. Yurchikhin set up a portable repress tank and replaced a pre-treat container in the Russian segment environmental control and life support system.

Yurchikhin and Misurkin carried out a fit check on their Orlan spacesuits on August 14, performed a preliminary suit leak check and practiced emergency procedures in the event of a pressure leak inside Pirs. Cassidy checked out the U. S. spacesuits after the water leak on Parmitano's suit during the July 16 spacewalk. Meanwhile, Parmitano worked throughout the day on cargo transfers in and out of the JAXA HTV-4. He rearranged stowage equipment in the HTV-4 and began filling it with trash. Nyberg helped in the morning and then conducted an interview with the NBC *Today* show. She also set up a combustion chamber inside the Multipurpose Small Payload Rack (MSPR) in the Kibō laboratory and replaced a carbon dioxide sensor. Vinogradov worked in the Russian segment of the station on various maintenance tasks. He carried out some plumbing work transferring urine to the ATV and cleaning vents and screens in the Zarya module and finally moved on to updating the station's inventory management system.

Cassidy and Parmitano continued removing cargo, reconfiguring packing equipment and filling the HTV-4 with trash on August 15. Nyberg and Cassidy set up a sensor and cleaned ventilation grilles to measure the air velocity inside Kibō. Nyberg then worked on a Japanese materials science experiment, Hicari, removing and replacing sample cartridges. Hicari takes place in the Kibō laboratory and uses a gradient heat furnace to produce high quality crystals that may advance the development of more efficient solar cells and semiconductor electronics. Cassidy also configured and installed a GLACIER science freezer into an EXPRESS rack inside the Destiny laboratory.

Yurchikhin and Misurkin began the 172nd spacewalk in support of station assembly and maintenance at 10:36 a.m. on August 16. They established the Strela cargo boom on the Poisk mini-research module. Misurkin operated the boom and manoeuvred Yurchikhin to the Zarya module near the Unity node. Yurchikhin then reran a cable connector and installed cables on Zarya. Misurkin installed the Vinoslivost experiment on Poisk. This is similar to the MISSE in that it exposes materials to the space environment. He then installed two connector patch panels and gap spanners on Poisk. Misurkin then helped Yurchikhin to install Ethernet cables on the Zarya cargo module. Unfortunately, this work was preparing for the arrival of the Nauka module, which has yet to be launched.

As is usual during spacewalks from the Russian segment, Vinogradov and Cassidy stayed in the Poisk module close to their Soyuz TMA-08M spacecraft while Nyberg and Parmitano stayed in the U. S. segment.

The hatch of the Pirs was closed and the spacewalk ended at 6:05 p.m. EDT. At 7 hours, 29 minutes, this was the longest Russian spacewalk to date.[12]

Expedition 36, Week Ending August 25, 2013

After the first of two spacewalks, Yurchikhin and Misurkin reviewed procedures and installing equipment on their Orlan spacesuits on August 19 to get ready for the second EVA scheduled for August 22. They participated in a debriefing with Russian spacewalk specialists to discuss Friday's spacewalk. Vinogradov, focused on maintenance tasks in the Russian segment as he cleaned fan screens in the Rassvet module and looked after the life-support system in the Zvezda service module. He worked on the Uragan experiment. This attempts to predict natural and man-made catastrophes, such as hurricanes. Nyberg removed the Nano Step experiment from the Solution Crystallization Observation Facility and replaced it with the Ice Crystal 2 experiment. Then she prepared new test samples for the Advanced Colloids Experiment (ACE), installed in the Light Microscopy Module inside the Fluids Integrated Rack. Cassidy and Parmitano weighed themselves with the Space Linear Acceleration Mass Measurement Device (SLAMMD), a device required as a weigh scale will not work in microgravity, SLAMMD generates a known force against an astronaut attached to an extension arm and calculates his or her body mass using Newton's Second Law. Parmitano then moved on to routine environmental health monitoring work as he measured sound levels throughout the orbiting complex. He rounded out his day supporting the InSpace-3 experiment. This studies colloidal fluids, or smart materials, that transition to a solid-like state in the presence of a magnetic field. Cassidy meanwhile set up three of the SPHERES in the Kibō laboratory. He was checking out a new graphical user interface to enable human-supervised control of satellites. Cassidy also observed another free-flying object outside the station near a docked Progress cargo vehicle. He reported the unidentified item to Mission Control Houston and captured video of it. It was later identified it as an antenna cover from the Zvezda service module.

On August 20, after the daily planning conference, Cassidy collected water samples from a condensate line in the Tranquility. Meanwhile, Parmitano cleaned the crew quarters, including the air intake and exhaust ducts and fan and airflow

---

[12] It was surpassed as the longest Russian spacewalk on December 27, 2013, by Oleg Kotov and Sergei Ryazansky who spent 8 hours, 7 minutes on their excursion. At the time of writing (July 2019) the longest Russian spacewalk was that of Alexander Misurkin and Anton Shkaplerov on February 2, 2018, lasting 8 hours, 13 minutes.

sensors. Then he performed an ultrasound on Cassidy for the Spinal Ultrasound investigation. Cassidy spoke to reporters from the Military Times and WBUR-FM in Boston. Nyberg replaced the infrared imager of the Fluid Physics Experiment Facility (FPEF) in the Ryutai rack in Kibō. Nyberg performed the Surface Telerobotics experiment, remotely controlling the K10 rover, which was at the Ames Research Center, Mountain View, California. In the Russian segment, Yurchikhin and Misurkin continued their preparations for a spacewalk by recharging the batteries of their Orlan spacesuits, checked out communication systems and prepared tools. Vinogradov conducted an audit of network and computer hardware and performed routine maintenance on the life support system inside the Zvezda service module.

Cassidy began August 21 by replacing the filters of the Water Recovery System. He then carried out an ultrasound test on Parmitano for the Spinal Ultrasound experiment. Cassidy, Misurkin and Vinogradov tested the "Kazbek" seat liners of their Soyuz TMA-08M spacecraft in preparation for their return to Earth. Nyberg set up commercial payloads in Kibō and replaced a rope on the Advanced Resistive Exercise Device (ARED). Parmitano replaced a sample cartridge in the Solidification and Quench Furnace. As part of the Crew Earth Observations program, the crew photographed wildfires in Idaho in the United States.

On August 22, Yurchikhin and Misurkin conducted their second spacewalk of the week. They opened the Pirs docking compartment at 7:34 a.m. They installed their combination EVA workstation and biaxial pointing platform at a temporary stowage location. Then they removed the External Onboard Laser Communications System, which was installed on Zvezda in August 2011 during an Expedition 28 spacewalk. When they started to install the new camera platform, they noticed that the base plate was not properly aligned. The advice of the ground controllers was to continue with the installation and that the platform could be aligned later. They installed it on the starboard side of the Zvezda module.

Misurkin and Yurchikhin then inspected six navigation antennas at various locations on the Zvezda module. These antennas provide navigation data during the rendezvous and docking of European ATV cargo ships. They tightened screws on loose antenna covers. A cover was observed to float away earlier so the Russian ground controllers were keen to check the other covers. Yurchikhin installed aids to future spacewalkers; handles on the exterior of the module. Misurkin collected a particulate sample from thermal insulation near Poisk's hatch and he took photographs of a materials exposure experiment. Ground controllers deferred the moving of a foot restraint to a future spacewalk They re-entered the station after 5 hours, 58 minutes at 1:32 p.m. EDT on Thursday. This was the 173rd spacewalk in support of space station assembly and maintenance.

Nyberg continued to work on the InSpace-3 experiment. Parmitano moved the exhaust port of the Amine Swingbed, an experiment to test more efficient carbon dioxide removal methods.

Yurchikhin and Misurkin spent August 23 drying out their Orlan spacesuits and participated in a debriefing with specialists at the Russian Mission Control Center in Korolev outside Moscow to discuss the spacewalk. Vinogradov and Parmitano re-opened the hatch to the ATV-4, which had been closed to support the space-walk. Nyberg continued work with the InSpace-3 experiment. She also set up the Combustion Integrated Rack for another round of experiments studying the process of combustion in microgravity. Cassidy performed a water recharge of the U. S. spacesuits in the Quest airlock, and he stowed tools that had been loaned to Yurchikhin and Misurkin for their recent spacewalks. Parmitano installed an improved Ethernet hub gateway to support science payloads inside the Columbus module. Throughout the day Parmitano and Cassidy also cooperated on a number of tasks as they tested air and surface samples throughout the station for microbial contamination and performed regularly scheduled maintenance on the Water Recovery System.

The robotics team at Houston's Mission Control Center started to move the Mobile Transporter railcar in order to relocate the Canadarm2 robotic arm. This was in preparation for the upcoming ground-commanded robotic movement of spare parts from the Exposed Pallet of the Kibō module to locations on the truss of the station.

Expedition 36, Week Ending September 1, 2013

On August 26, after regular health evaluations, Nyberg continued working on the InSPACE-3 experiment. Cassidy charged batteries for a technology demonstration of the SPHERES. Parmitano set up the Combustion Integrated Rack and reviewed the plans to investigate the Biolab's microscope cassette. Biolab is situated in the Columbus module and biological experiments are conducted there. Later, he collected and tested station's potable water samples for contamination. Cassidy, Vinogradov and Misurkin prepared for their journey home aboard the Soyuz TMA-08M spacecraft. Vinogradov also refilled water containers from the storage tanks aboard the ATV-4. Before HTV-4 left the station, Nyberg stowed items for disposal when the spacecraft burns up on re-entry. Cassidy removed light bulbs for re-use aboard the station.

The robotics team at Mission Control Center stowed the Special Purpose Dexterous Manipulator and moved Canadarm2 to a new location. They used the Canadarm2 to remove a spare Main Bus Switching Unit and a spare Utility Transfer Assembly from the Exposed Pallet and attach the spares to the station's truss.

Parmitano stowed items for disposal aboard the Orbital Sciences Cygnus cargo craft on August 27. It was scheduled to launch from the Wallops Flight Facility, Virginia, in the United States, on September 17. Nyberg set up two SPHERES to test the Resonant Inductive Near-field Generation System (RINGS) experiment, where power is transferred between two satellites without physical contact. Cassidy and Parmitano checked out the spacesuit that Parmitano wore during the aborted July spacewalk. They powered it up and observed that the helmet began again to fill with water. With the issue replicated, mission controllers started planning to replace parts one by one to identify the source of the problem. Cassidy, Vinogradov and Misurkin participated in a call with the search and rescue team that would meet them after their upcoming landing in Kazakhstan. After the teleconference, Vinogradov, Misurkin and Yurchikhin had their cardiac bioelectric activity measured. Vinogradov also spent some time conducting the Seiner ocean-observation experiment.

Mission controllers in Houston powered up the Special Purpose Dexterous Manipulator (Dextre) and used it to move the Space Test Project-4 payload from the Exposed Pallet at the forward end of the Kibō module to the station's truss.

Cassidy powered up Robonaut on August 28 so that ground controllers commanded it to remove and manipulate a sample insulation cover typical of the ones found aboard the station. This was with a view to Robonaut replacing astronauts on hazardous spacewalks. He also installed the FROST freezer inside the Japanese Kibō module. Nyberg and Parmitano carried out tonometry tests to measure the intraocular pressure in their eyes. Afterward, Parmitano performed troubleshooting on Biolab's microscope cassette in the Columbus module. He then removed and stowed slow growth samples from the Binary Colloid Alloy Test experiment. Nyberg meanwhile collected air and surface samples throughout the complex and tested them for signs of microbial contamination. She also packed up some of the samples for return to Earth aboard the Soyuz TMA-08M spacecraft. Nyberg talked with TV host Arsenio Hall of the CW. Vinogradov wore the Lower Body Negative Pressure outfit to simulate gravity's effect on the body by drawing the body's fluids to the lower half. This was in preparation for his return to Earth. Misurkin and Yurchikhin continued to stow tools and equipment used during their two recent August spacewalks. The crew also photographed the Mississippi delta, widespread flooding in Pakistan and the Rim Fire around California's Yosemite National Park and the Stanislaus National Forest.

Overnight, the robotics team at Mission Control Center removed the Space Test Project-3 payload from the station's truss and placed it on the Exposed Pallet.

On August 29, Cassidy repeated his Robonaut experiment of the previous day. The exception was that this time, the humanoid robot was mimicking Cassidy's movements rather than being remotely controlled from the ground. Once completed and stowed, he then gathered items for disposal aboard the soon to launch Orbital Sciences Cygnus cargo craft. Meanwhile, Nyberg prepared tools in order to replace a water relief valve inside the spacesuit that failed during Parmitano's EVA. Parmitano, Nyberg and Yurchikhin carried out a periodic emergency drill by reviewing their roles and responsibilities for an emergency descent aboard their Soyuz TMA-09M spacecraft. At the end of the day, Cassidy used a fundoscope to examine the eyes of Parmitano and Nyberg.

Nyberg and Parmitano began August 30 with a battery of medical tests including ultrasound examination of their eyes, blood pressure measurements and cardiac scans for the Ocular Health study. Cassidy worked on the eValuatIon And monitoring of microBiofiLms insidE the ISS (VIABLE).[13] This evaluates microbial biofilm development on space materials. Afterward, Cassidy and Nyberg prepared for future investigation of Parmitano's spacesuit after the EVA water leak. Vinogradov, Misurkin and Cassidy performed air-leak tests on their Sokol launch/entry suits before their upcoming return to Earth.

The robotics ground controllers in Houston, with help from Cassidy and Parmitano, used the Canadarm2 to place the HTV-4's Exposed Pallet and an attached Department of Défense payload back into its unpressurized segment.

Nyberg commenced the Asian Seed experiment by watering pots containing azuki bean seeds. Students in Asia participated in this educational experiment to learn about the importance of space biology as they compared plant growth in space to that on Earth. Misurkin collected data from the Matryoshka experiment in the Russian segment. Meanwhile, Yurchikhin replaced dust collector filters in the Zarya module and maintained life-support systems in the Zvezda module.

The ATV Albert Einstein was used again to raise the station's orbit on August 31 by firing its engines for 3.5 minutes. This prepared the station for the departure and arrival of Soyuz spacecraft.

Expedition 36, Week Ending September 8, 2013

On September 3 Parmitano and Nyberg closed the hatches between the station and the JAXA HTV-4 cargo vehicle. They installed controller panel assemblies so that Mission Control could remotely prepare the Harmony's Common Berthing Mechanism for the HTV-4's unberthing. Vinogradov, Cassidy and Misurkin, meanwhile, prepared their crewed Soyuz spacecraft for its upcoming return trip. Although the HTV is designed to burn up over the Pacific Ocean, the Soyuz can return the crew and limited cargo safely to Earth.

After the ground controllers unberthed the HTV-4 from the nadir port of the Harmony module on September 4, Nyberg took control of the Canadarm2 and released the spacecraft at 12:20 p.m. EDT. Meanwhile, Vinogradov, Misurkin and Cassidy continued preparations by participating in onboard Soyuz descent training and conducting an equipment and stowage briefing.

On September 5, Nyberg checked the insulation inside a gradient heater furnace in the Kibō laboratory and videotaped plants cultivated for a Japanese botany experiment to encourage students to study science. Cassidy and Parmitano used smartphones to control the SPHERES satellites, while Yurchikhin carried out maintenance tasks, including copying data for the Identification study that documents dynamic loads on the space station and measuring radiation in the Russian segment for the Matroyshka experiment.

---

[13] Another contender for most tortured acronym award.

Vinogradov, Cassidy and Misurkin finalized their Soyuz descent training on September 6. Cassidy photographed samples collected for the Binary Colloidal Alloy Test (BCAT-C1) experiment. Nyberg and Parmitano took spinal scans using ultrasound and electrocardiogram equipment. Cassidy prepared cameras in the ESA cupola to photograph the re-entry of the HTV-4.

Expedition 36/37, Week Ending September 15, 2013

At 2:25 p.m. EDT on September 9, Expedition 36 Commander Pavel Vinogradov handed over control of the International Space to Fyodor Yurchikhin of Expedition 37. Expedition 37 was scheduled to begin when the Soyuz carrying Vinogradov, Cassidy and Misurkin undocked the following day. This was Yurchikhin's fourth space mission and mission on the ISS. He was a member of Expedition 15 in 2007 and Expeditions 24/25 in 2010.[14] Before departing, Cassidy collected blood and urine samples for stowage inside the Human Research Facility's science freezer. He also replaced a fluids control and pump assembly inside the Tranquility node's Water Recycling System. Vinogradov and Misurkin prepared for departure by reviewing the Soyuz undocking procedures. Vinogradov stowed the last equipment in the Soyuz while Misurkin practiced Soyuz descent operations. Nyberg and Parmitano, who were remaining on the station, reviewed rendezvous procedures and prepared for the arrival of the Cygnus, expected later in September. They practiced robot grappling operations.

Soyuz TMA-08M, carrying Expedition 36 crew members Pavel Vinogradov, Chris Cassidy and Alexander Misurkin, undocked from the Poisk mini-research module at 7:37 p.m. EDT on September 10, ending a five-and-a-half month stay at the International Space Station. They landed in Kazakhstan at 10:58 p.m.

Reduced to a crew of three, Expedition 37's Yurchikhin, Parmitano and Nyberg awaited the arrival of Oleg Kotov, Sergei Ryazansky and Michael Hopkins, who were scheduled to be launched on September 25.

September 11 was a rest day for the crew of Expedition 37, after a busy scheduling monitoring and photographing the departing Soyuz TMA-08M.

Yurchikhin checked out the Elektron oxygen generation system in the Russian segment on September 12, purging the device after shutting it down for maintenance. In the afternoon, he worked on the Matroyshka radiation exposure experiment. Parmitano worked on science inside the Destiny laboratory's Microgravity Science Glovebox, conducting another run of the long-running materials science experiment, InSPACE-3. Nyberg worked throughout her morning cleaning the crew quarters in the overhead section of the Harmony node. She partially disassembled the crew quarters and cleaned its panels, exhaust ducts, fans and air sensors. Parmitano and Nyberg packed cargo to be stowed inside Orbital Sciences' Cygnus spacecraft.

---

[14] In 2017, he would go on to be a crewmate of ESA's Thomas Pesquet during Expedition 51 and to command Expedition 52, with ESA's Paolo Nespoli also aboard.

Yurchikhin worked with spacewalk tools on September 13; he copied data from the Identification experiment to a laptop computer, and he checked the ventilation system in the Zvezda service module, examining pressure and atmosphere sensors. Nyberg collected data from the InSPACE materials science experiment in the Destiny laboratory's Microgravity Science Glovebox. Parmitano replaced a mass spectrometer in the Major Constituent Analyzer (MCA) in the Tranquility module. The MCA records nitrogen, oxygen, carbon dioxide, methane, hydrogen and water vapor levels.

Parmitano and Nyberg practiced robotics procedures for the capture and berthing of the first Orbital Sciences' Cygnus resupply craft, due late in the month.

The thrusters of the ATV-4 were fired for 3 minutes, 25 s beginning at 8:42 a.m. August 15. The reboost raised the perigee of the station's orbit by one mile and left the complex in an orbit of 262.7 × 253.7 statute miles, ready for the arrival of Soyuz TMA-10M.

Expedition 37, Week Ending September 22, 2013

Nyberg started the week of September 16 setting up the Resist Tubule experiment in the Saibo rack of the Kibō module. This examines the mechanisms of gravity resistance in plants. It should assist researchers learn more about the evolution of plants and enable efficient plant production not only here on Earth but in space as well. Afterwards, she replaced emergency mask kits and updated their procedure books. She reviewed these procedures with Parmitano and Yurchikhin, for use in the event of an ammonia leak. Parmitano spent the remainder of his morning conducting Crew Medical Officer training and checking air and water samples for signs of contamination. Nyberg spoke with Marcia Dunn of the Associated Press about life aboard the station. Nyberg and Parmitano used the Robotics OnBoard Trainer (ROBoT) to review the procedures for grappling Cygnus with the Canadarm2. Parmitano also installed the Common Berthing Mechanism (CBM) Centerline Berthing Camera System (CBCS) inside Harmony to help the flight control teams monitor the activities. He also gathered and verified the hardware needed to outfit the vestibule where Cygnus was planned to be berthed. Yurchikhin worked with the Kulonovskiy Kristall experiment and the Matryoshka experiment.

Nyberg began her workday on September 17 by setting up a new test sample for the Advanced Colloids Experiment (ACE) in the Light Microscopy Module in the Fluids Integrated Rack. Yurchikhin and Parmitano meanwhile conducted a review of emergency descent procedures for their Soyuz TMA-09M spacecraft. Nyberg and Parmitano continued the Ocular Health, measuring their blood pressure and testing the intraocular pressure of their eyes using a tonometer. After a break for lunch, Nyberg and Parmitano used the robotics workstation inside the station's cupola to practice techniques for grappling Orbital Sciences' Cygnus craft. They grappled the fixture on the *Leonardo*

Multi-Purpose Logistics Module to rehearse the robotic capture of Cygnus in an offset position. Yurchikhin worked on the Kulonovskiy Kristall experiment.

He also checked a number of panels in the Zvezda service module to see if they were being affected by vibrations from the treadmill.

The first Orbital Sciences Cygnus cargo spacecraft destined to berth at the ISS was launched atop an Antares rocket at 10:58 a.m. EDT, September 18, 2013, from the Mid-Atlantic Regional Spaceport Pad-0A at NASA's Wallops Flight Facility in Virginia. At the time of launch, the station was about 261 miles above the southern Indian Ocean. The Cygnus craft 'G. David Low'[15] was carrying 1300 pounds of cargo, including food and clothing.

The three Expedition 37 crew members watched the launch online from the Destiny laboratory. Nyberg and Parmitano then reviewed the Cygnus' cargo manifest and planned its unloading. The three of them had training to review their roles and responsibilities in the event of an emergency aboard the station such as a fire or rapid depressurization.

Parmitano continued the Pro K experiment on September 19 to evaluate the effectiveness of dietary changes to reduce the bone loss experienced by astronauts while in microgravity. Nyberg configured the Environmental Research and Visualization System (ISS SERVIR) camera in the Window Observational Research Facility (WORF), in the Destiny laboratory. This system provides researchers the means to gain experience and expertise in automated data acquisition from the station as they direct the camera to collect imagery of specific areas of the world for disaster analysis and environmental studies. Nyberg assisted the ground team on Thursday with efforts to fine-tune the alignment and pointing motion of the camera. She then configured equipment for the robotics workstation inside the station's cupola to support the arrival of the Cygnus. Station commander Yurchikhin worked with science programs such as the Kulonovskiy Kristall experiment, the micro-accelerometer for the Identification experiment and the Uragan Earth-observation experiment.

Early on September 20 Nyberg and Parmitano planned the rendezvous of the Cygnus approach, then reviewed the manifest and the grapple and berthing procedures. They also participated in several ongoing medical studies such as the Pro K diet logging program. Yurchikhin worked on the Kulonovskiy Kristall experiment, then cleaned air ducts in the Poisk Mini-Research Module and maintained the life-support system in the Zvezda service module.

Expedition 37, Week Ending September 29, 2013

Parmitano worked with the InSPACE-3 experiment in Microgravity Science Glovebox in the Destiny laboratory on September 23. He also participated in the Skin B experiment, which investigates skin aging mechanisms that are slow on Earth but very much accelerated in weightlessness. Nyberg replaced a manifold bottle in the Combustion Integrated Rack. This equipment rack contains an optics

---

[15] G. David Low was a NASA astronaut who flew on STS-32, STS-43 and STS-57. He worked for Orbital Sciences after retiring from NASA.

bench, combustion chamber, fuel and oxidizer control and five cameras. She also replaced filters in the Tranquility node, which houses many of the space station's life support and environmental control systems. Yurchikhin conducted another session of the Coulomb Crystal experiment, investigating charged particles in microgravity. He also collected dosimeter radiation readings for the Matryoshka experiment.

Three new members of Expedition 37 launched from Baikonur Cosmodrome, Kazakhstan at 4:58 p.m. on September 25. Oleg Kotov was on his third flight to the station, while Mike Hopkins and Sergey Ryazanskiy were on their first space-flights. The Soyuz TMA-10M spacecraft docked at the Poisk mini-research module at 10:45 p.m. EDT and the hatches opened at 12:34 a.m. EDT. They were greeted by Expedition 37 Commander Fyodor Yurchikhin and Flight Engineers Karen Nyberg and Luca Parmitano.

After talking with families, the six-member crew shared a meal and got to work. The two new cosmonauts performed some light cargo transfer work and Hopkins filled out a health questionnaire. Then the entire crew gathered for a safety briefing to review emergency procedures and equipment locations. They also familiarized themselves with the potential hazards and available safety measures aboard the station. September 26 was an off-duty day for the crew as they sleep shifted back to their normal schedules. Yurchikhin, Nyberg and Parmitano had stayed up for several hours after their bedtime to greet their new crewmates and monitor their arrival.

This was followed by a day of light duty on September 27, during which the new arrival unpacked their Soyuz TMA-10M spacecraft, while the others performed light maintenance and exercised. Parmitano celebrated his birthday.

The first Orbital Sciences Cygnus cargo spacecraft was captured by the station's Canadarm2 at 7 a.m. EDT on September 29. After a computer data link issue, the approach was postponed by a week. Orbital Sciences uploaded a software fix and then it was decided to wait until after the Soyuz had docked. Parmitano was at the controls of the Canadarm2 in the cupola with Nyberg as a backup at the secondary robotics workstation in the Destiny laboratory. Cygnus aimed a laser at a reflector on the Kibō laboratory, fulfilling its final requirement. It then moved to its capture point about 10 m from the station. The Cygnus spacecraft drifted and Parmitano used the Canadarm2 to grapple the Cygnus. He attached it to the nadir port of the Harmony node at 8:44 a.m.

Expedition 37, Week Ending October 6, 2013

Nyberg and Parmitano opened the hatch to the Cygnus at 4:10 a.m. EDT on September 30 and began unloading its 1300 pounds of cargo. Parmitano and Hopkins replaced a water line vent tube assembly on the U. S. spacesuit that Parmitano wore on his previous spacewalk. Yurchikhin worked on the Aseptic experiment, which takes a look at the methods and means of ensuring sterile conditions for biotechnological experiments aboard the station. Ryazanskiy assisted the commander with photography of the experiment results. Ryazanskiy also performed the Kaskad investigation to help researchers learn how to increase the output of target bioactive substances in the cultivation of cells under microgravity conditions. Kotov meanwhile focused on the Konstanta experiment, which studies enzyme reactions in a microgravity environment. Afterward, he unloaded cargo from the Soyuz TMA-10M spacecraft docked to the Poisk module.

From October 1 to 17, the U. S. federal government was shut down, with non-essential routine tasks halted and staff furloughed. While the American crew on the ISS continued their tasks, the NASA staff who prepared the mission reports did not. During this time, routine tasks were performed, and the ATV-4 was again used to boost the station's altitude with an 818 second burn on October 1. Parmitano conducted the 10-day ENERGY experiment, wearing a tricep armband to measure body mass balance. Hopkins completed a 3 days ocular health survey, ultrasound scans and blood pressure monitoring. Nyberg celebrated her birthday on October 7.

Expedition 37, Week Ending October 20, 2013

Parmitano and Nyberg spent the October 17 training and reviewing procedures for departure of the Cygnus spacecraft. Nyberg and Hopkins also finished loading trash onto the craft. Kotov installed and connected a control panel to monitor the departure of the ESA ATV-4. Hopkins performed an ultrasound on Nyberg for the Spinal Ultrasound investigation.

On October 17, Parmitano spoke with more than 50 students from five countries participating in the Volare Space Robotics challenge at ESTEC in the Netherlands. The teams used robots to unload cargo from a model ATV. He told them "Students represent our future, and you should be proud of what you accomplished."

Parmitano conducted the Canadian BP Reg experiment. This is a medical study to understand fainting and dizziness experienced by some astronauts after they return to Earth. Nyberg set up the Combustion Integrated Rack to continue to evaluate combustion in microgravity. Then she conducted a robotics test session with two SPHERES in the Japanese Kibō laboratory. Hopkins drew blood for the Human Research Facility, completed a proficiency training session in his role as crew medical officer and collected surface samples throughout the U. S. segment of the station and incubated them to look for signs of microbial contamination. Hopkins and Nyberg spoke with Tim Sherno of St. Paul's KSTP-TV in Nyberg's home state of Minnesota and the Big Ten Network anchor Rick Pizzo. Hopkins talked with former NFL fullback Howard Griffith, who was one of Hopkins' Fighting Illini teammates at the University of Illinois. Hopkins and Parmitano stowed trash aboard the ATV-4, which was destined to deorbit 10 days later. Kotov and Ryazanskiy studied spacewalk procedures and reviewed a preliminary timeline for an upcoming Russian EVA. Ryazanskiy also manually mixed samples for the Cascade biotechnology experiment, which investigates cell cultivation in weightlessness.

Expedition 37, Week Ending October 27, 2013

The hatches between Cygnus and the station were closed Monday at 6:42 a.m. EDT, October 21. Nyberg, Parmitano and Hopkins configured the Harmony node's vestibule for Cygnus' demating and depressurizing. Afterwards, the trio gathered to review their Cygnus departure activities.

Parmitano and Nyberg controlled the Canadarm2 when it released the Orbital Sciences' Cygnus cargo craft at 7:30 a.m. EDT, October 22 after three weeks docked at the station. This was the first Cygnus to berth at the station and was the demo flight before commercial activities commenced.

Nyberg analyzed water samples, checked a science experiment and collected blood samples for storage in a science freezer. Parmitano worked on a couple of science experiments and took photos inside of Europe's Albert Einstein ATV-4. Hopkins also gathered U. S. spacesuit tools and worked on a fluid physics experiment. Yurchikhin worked on Russian maintenance tasks while Ryazanskiy and Kotov worked throughout the morning gathering spacewalk tools for the upcoming Russian spacewalk.

The Cygnus was deorbited over the Pacific Ocean on October 23. It was a day for spacesuit maintenance on the station. Nyberg and Hopkins continued troubleshooting the U. S. spacesuit water leak scrubbing cooling loops and preparing to replace a fan pump separator, while Kotov and Ryazanskiy readied their Russian spacesuit equipment for an upcoming spacewalk. Parmitano worked on science throughout the day. He measured his forearm for a skin-aging experiment. He later photographed samples for the Binary Colloidal Alloy Test-C1 experiment, which studies nano-scale particles dispersed in liquid. Finally, Parmitano sampled the station's surfaces and air for microbes and then changed a lens for an Earth observation camera remotely controlled by students. Yurchikhin worked in the station's Russian segment on various science and maintenance tasks. He participated in an experiment studying the veins in the lower extremities of a cosmonaut. He then worked on the Zvezda service module's ventilation system and checked flow sensor positions

The ATV-4, Albert Einstein, fired its engines to raise the station's orbit for the last time on October 24. Parmitano talked to students and officials from Italy, Germany and Israel to commemorate the ATV-4. Nyberg and Hopkins continued to troubleshoot the faulty U. S. spacesuit by removing and replacing a fan pump separator, inspecting and photographing their work and discussing their findings with ground controllers. Kotov and Ryazanskiy also worked on spacesuits, preparing their Orlan suits for an upcoming November spacewalk. They conducted leak checks inside the Pirs docking compartment and recharged water systems and batteries inside their spacesuits. Yurchikhin set up the Iridium phone to be installed in the Soyuz TMA-09M. Nyberg worked on the Japanese Biological Rhythms 48 study while Parmitano participated in the European Circadian Rhythms investigation.

Parmitano and Kotov closed the ATV-4's hatches on October 25. Parmitano then joined Yurchikhin in a Soyuz departure conference with search and rescue team specialists.

Expedition 37, Week Ending November 3, 2013

The ESA Automated Transfer Vehicle (ATV-4) Albert Einstein undocked from the aft port of the Zvezda service module at 4:55 a.m. EDT on October 28, monitored by Parmitano and Kotov who were in the Zvezda module. They could have taken control, if required. Nyberg installed a dehumidifier in the Cell Biology Experiment Facility. She also performed an ultrasound on Flight Engineer Mike Hopkins for the Spinal Ultrasound investigation. Astronauts can grow up to 3% taller during long duration missions. They always revert to their normal height eventually. The portable Spinal Ultrasound equipment is used to investigate this phenomenon. Afterward, Hopkins gathered and inspected spacewalking tools and tethers.

In the Russian segment of the station, Flight Engineer Sergey Ryazanskiy tested the transmission of images through slow-scan television via the station's amateur radio equipment. He also maintained the life-support system in the Zvezda module.

Yurchikhin and Parmitano conducted an onboard descent drill inside the Soyuz TMA-09M on October 29. They reviewed emergency scenarios and spoke with flight controllers to coordinate the list of items being returned aboard the Soyuz. Hopkins replaced an air selector valve within the Carbon Dioxide Removal Assembly in the Tranquility module. Nyberg monitored the Ice Crystal 2 experiment to assist the ground team with troubleshooting an issue with the payload's temperature control. She also participated in the Microbiome study. This investigates the impact of space travel on the human immune system and the microorganisms that reside in our bodies.

Yurchikhin and Parmitano made preparations ahead of relocating their Soyuz TMA-09M spacecraft from the Rassvet module to the aft port of the Zvezda service module. This was in order to free up the Rassvet port for the arrival of Soyuz TMA-11M.[16]

Yurchikhin also packed equipment for the return to Earth and participated in the Lower Body Negative Pressure test. Parmitano, Nyberg and Hopkins participated in a battery of medical checkups and experiments as researchers kept track of the crew's health and developed countermeasures to minimize or prevent the harmful effects of long-duration spaceflight. Hopkins and Nyberg followed the Pro K diet plan. They also tested urine samples and stored blood draws inside the

---

[16]Coincidentally, Yurchikhin piloted the previous Soyuz relocation in June 2010. Then, Expedition 24's Soyuz TMA-19 was moved from Zvezda to the then newly installed Rassvet.

Minus Eighty Degree Laboratory Freezer for ISS (MELFI), to preserve them for additional analysis back on Earth.

To insure their cardiovascular and musculoskeletal health and performance remained satisfactory, Parmitano and Hopkins each donned blood pressure cuffs and electrocardiograph monitors as they performed graded exercises on a stationary bicycle. Hopkins also performed an ultrasound on Parmitano for the Spinal Ultrasound investigation. Nyberg, Hopkins and Parmitano later weighed themselves with the Space Linear Acceleration Mass Measurement Device (SLAMMD). Nyberg also retrieved and stored samples inside MELFI for the Resist Tubule experiment. Nyberg rounded out her day cleaning the exhaust ducts, fans and airflow sensors in the starboard crew quarters.

Expedition 37, Week Ending November 10, 2013

Yurchikhin, Nyberg and Parmitano undocked their Soyuz TMA-09M spacecraft from the nadir port of the Rassvet module at 4:33 a.m. EDT November 1. The moved away from the station, rotated and began the flyaround to the rear of the station. Yurchikhin guided the spacecraft to dock at the aft port of the Zvezda module at 4:54 a.m. This had recently been the location of the ATV-4. The move would allow the Soyuz TMA-11M spacecraft to dock at Rassvet in the coming week.

At the same time, Kotov and Ryazanskiy prepared their tool caddies for their upcoming spacewalk. Hopkins set up the Advanced Biological Research System (ABRS). He also recharged batteries for the SPHERES.

Hopkins worked with the SPHERES-RINGS experiment on November 4, to study relative station-keeping, maneuvering, and attitude control between two satellites. Yurichikhin, Parmitano and Nyberg packed for their trip home. Included in their cargo were parts from the leaking U. S. spacesuit for inspection by engineers on the ground.

On November 5, Hopkins collected water samples and transferred fluids through a condensate pump. He spoke with the Weather Channel and KSDK-TV from St. Louis, worked on a treadmill for its monthly and yearly maintenance and worked with the InSPACE experiment. Nyberg conducted a run of the ongoing SPHERES experiment using student-written algorithms to control the small satellites. She and Parmitano partnered up for eye exams with remote guidance from the ground. Yurchikhin stowed equipment and trash inside the Soyuz.

Hopkins participated in the Body Measures experiment, on November 6. This collects anthropometric data to help researchers understand the magnitude and variability of the changes to body measurements during spaceflight. Nyberg assisted, setting up the calibration tape, collecting data and taking photographs. Parmitano completed his session with Biological Rhythms 48hrs, a JAXA study of the circadian variation of astronauts' cardiac function during spaceflight using a small digital electrocardiograph. Later, Parmitano, Nyberg and Yurchikhin carried out a descent drill as they prepared for their departure. Yurchikhin, Kotov and Ryazanskiy also reviewed hatch opening procedures in anticipation of the new arrivals, NASA's Rick Mastracchio, JAXA's Koichi Wakata and Mikhail Tyurin of Roscosmos.

Tyurin, Wakata and Mastracchio launched from Baikonur Cosmodrome, Kazakhstan, at 11:14 p.m. November 7 aboard Soyuz TMA-11M. They docked to the Rassvet docking compartment at 5:27 a.m., and the hatches were opened at 7:44 a.m. EST. With nine crew members onboard, this was the first time since October 2009 that nine people had been on the station since the retirement of the space shuttle. It was an experienced trio. This was Tyurin's third mission to the ISS, Mastracchio's fourth and Wakata's third, although Wakata also flew on a shuttle mission that did not visit the ISS. They brought the Olympic torch of the 2014 Winter Olympic Games at Sochi, Russia. It was to be carried outside the station on a Russian EVA.

On November 8, all nine crewmembers participated in a news conference in the Destiny laboratory. Parmitano worked inside the Columbus laboratory module swapping sample cassettes and repairing a BioLab microscope. Nyberg and Hopkins installed a wireless access point inside the Kibō laboratory. Later, Hopkins collected station water samples for microbial analysis. Nyberg spent her afternoon packing for her return home as well as continuing handover activities with the next crew. Kotov and Ryazanskiy prepared for a planned 6 hours spacewalk.

Kotov and Ryazanskiy opened the hatch to the Pirs docking compartment at 9:34 a.m. EST, November 9, and carried the unlit Olympic torch outside the station.[17] After photographing the torch, they stowed it in the airlock and moved out to the hull of the Zvezda service module. They set up the EVA activity workstation and biaxial pointing platform that was installed during an Expedition 36 EVA in August. They installed handrails on the workstation, then loosened three bolts and removed a launch bracket from the pointing platform. They noticed the misalignment of foot restraint and hence deferred its installation. Finally, they deactivated the Radiometria experiment package on Zvezda. This device collected data for seismic forecasting and earthquake predictions, Radiometria had been set up and mounted during an Expedition 26 spacewalk in February 2011. Another task was deferred; folding down and tying Radiometria's antenna. The spacewalkers closed the Pirs hatch at 3:24 p.m.

This was the 174th spacewalk in support of space station assembly and maintenance. It was Kotov's fourth and Ryazanskiy's first.

---

[17]This was the third time an Olympic torch was launched into space but the first time it was taken on an EVA. The prior occasions were STS-78 in 1996 ahead of that year's Atlanta summer games and STS-101 in 2000 ahead of the Sydney summer games.

Fig. 3.10. Cosmonauts Oleg Kotov and Sergey Ryazanskiy carry the Sochi Olympic torch in space, November 2013 (Roscosmos)

On November 10 Yurchikhin handed command of the station to Kotov during a change of command ceremony. Yurchikin, Nyberg and Parmitano boarded the Soyuz and the hatches between the spacecraft were closed at 3:09 p.m. The Soyuz TMA-09M undocked from the aft port of the Zvezda at 6:26 p.m. EST Sunday as the station was 262 miles over northeast Mongolia. The undocking was the end of Expedition 37 and the start of Expedition 38. The Soyuz landed southeast of Dzhezkazgan at 9:49 p.m.

Postscript

In April 2014, Parmitano participated in the annual ESA Cooperative Adventure for Valuing and Exercising human behavior and performance Skills (CAVES) week-long training program, where astronauts from all the ISS participating space agencies explore the caves at Sa Grutta in Sardinia in order to "improve leadership, teamwork, decision-making and problem-solving skills". He was joined by ESA's Matthias Muarer, NASA's Scott Tingle and Alexander Misurkin and Sergey Kud-Sverchkov, both from Roscosmos.

In July 2015, Parmitano commanded the 20th NASA Extreme Environment Mission Operations (NEEMO) crew. This is another mission analog for astronauts where they live and work in the Florida International University (FIU) Aquarius underwater laboratory, 5.6 km off Key Largo, Florida, and 19 m below the surface.

In September and October 2016, Parmitano participated in the first ESA Pangaea program, where astronauts follow in the footsteps of the Apollo astronauts who studied geology in the field. Together with the CAVES program this should prepare astronauts for the next generation of missions to the Moon and Mars. The first Pangaea students were Luca Parmitano, Pedro Duque and Matthias Maurer, all from ESA.

Parmitano served as the lead EVA communicator during ESA astronaut Thomas Pesquet's first 5 hours and 58 minutes spacewalk on his Proxima mission on January 13, 2017.

ESA Director General Jan Wörner announced on January 18, 2017, that Luca would return to the ISS as part of Expedition 60/61 in 2019.

At the time of writing, July 2019, Parmitano has launched on Soyuz MS-13 to join Expedition 60. His mission is called Beyond, and he is to command Expedition 61, the third ESA astronaut and first Italian to command the station.

Fig. 3.11. Luca Parmitano training for the Beyond mission at the Gagarin Cosmonaut Training Center, Moscow, Russia, June 2019 (ESA)

During his first space mission, Parmitano accumulated 166 days, 6 hours and 17 minutes in space. At the time of this writing he has begun participating in his second mission, Beyond.

4

Blue Dot

Mission

ESA Mission Name: Blue Dot
Astronaut: Alexander Gerst
Mission Duration: 165 days, 8 hours, 1 minutes
Mission Sponsors: DLR/ESA
ISS Milestones: ISS 39S, 76th crewed mission to the ISS

Launch

Launch Date/Time: May 28, 2014, 19:57 UTC
Launch Site: Pad 1, Baikonur Cosmodrome, Kazakhstan
Launch Vehicle: Soyuz TMA
Launch Mission: Soyuz TMA-13M
Launch Vehicle Crew: Maksim Viktorovich Surayev (RKA), CDR
                      Gregory Reid Wiseman (NASA), Flight Engineer
                      Alexander Gerst (ESA) Flight Engineer

Docking

Soyuz TMA-13M
Docking Date/Time: May 29, 2014, 01:44 UTC
Undocking Date/Time: November 10, 2014, 00:31 UTC
Docking Port: Rassvet nadir

Landing

Landing Date/Time: November 10, 2014, 03:58 UTC
Landing Site: near Arkalyk, Kazakhstan
Landing Vehicle: Soyuz TMA
Landing Mission: Soyuz TMA-13M
Landing Vehicle Crew: Maksim Viktorovich Surayev (RKA), CDR
                      Gregory Reid Wiseman (NASA), Flight Engineer
                      Alexander Gerst (ESA) Flight Engineer

© Springer Nature Switzerland AG 2020
J. O'Sullivan, *European Missions to the International Space Station*,
Springer Praxis Books, https://doi.org/10.1007/978-3-030-30326-6_4

ISS Expeditions

ISS Expedition:	Expedition 40
ISS Crew:	Steven Ray Swanson (NASA), ISS-CDR
	Aleksandr Aleksandrovich Skvortsov (RKA), ISS-Flight Engineer 1
	Oleg Germanovich Artemyev (RKA), ISS-Flight Engineer 2
	Maksim Viktorovich Surayev (RKA), ISS-Flight Engineer 3
	Gregory Reid Wiseman (NASA), ISS-Flight Engineer 4
	Alexander Gerst (ESA), ISS-Flight Engineer 5
ISS Expedition:	Expedition 41
ISS Crew:	Maksim Viktorovich Surayev (RKA), ISS-CDR
	Gregory Reid Wiseman (NASA), ISS-Flight Engineer 1
	Alexander Gerst (ESA), ISS-Flight Engineer 2
	Aleksandr Mikhailovich Samokutyayev (RKA), ISS-Flight Engineer 3
	Yelena Olegovna Serova (RKA), ISS-Flight Engineer 4
	Barry Eugene Wilmore (NASA), ISS-Flight Engineer 5

The ISS Story So Far

The Chinese continued their gradual progress when another three-person crew entered the Tiangong-1 space station. The second female Chinese taikonaut, Wang Yaping, was on the crew of Shenzhou 10 in June 2013.

Soyuz TMA-10M launched in September 2013 to deliver crew members to Expedition 37.

In November 2013, Soyuz TMA-11M delivered Expedition 38/39 crew to the station, including Koichi Wakata who became the first Japanese commander of the ISS when he assumed command of Expedition 39 in March 2014.

Soyuz TMA-12M was expected to make the recently introduced, 4-orbit, 6 hours rendezvous, but when the third maneuver did not occur as expected Mission Control told the crew to revert to the older 2 days rendezvous, which led to a successful docking with the station on March 27, 2014.

Alexander Gerst

Alexander Gerst was born in Künzelsau, Baden-Württemberg Germany, on 3 May 1976. He graduated from the Technical High School in Öhringen, Germany, in 1995. During his school days, he volunteered as a scout leader, firefighter and lifeguard. He earned a degree with distinction in geophysics from the University of Karlsruhe, in 2000 and a master's degree in Earth sciences from the Victoria University of Wellington, New Zealand, also with distinction, in 2003. During his masters and doctoral studies, he participated in field experiments in Antarctica and New Zealand including monitoring volcanoes. As a result, he has had papers published in *Science* and *Nature* and won the German Research Foundation (DFG) Bernd Rendel award for outstanding research in 2007.

He worked as a researcher at the Institute of Geophysics from 2004 and was awarded a PhD in Natural Sciences from the University of Hamburg in 2010. His dissertation was on volcanic eruption dynamics.

Fig. 4.1. Alexander Gerst. (ESA)

He is a member of the following organizations:

- The International Association of Volcanology and Chemistry of Earth's Interior (IAVCEI)
- The German Geophysical Society (DGG)
- The European Geosciences Union (EGU)
- The European Volcanological Society (SVE)
- The American Geophysical Union (AGU)

Gerst is single and enjoys sports, such as fencing, swimming and running. He participates in many outdoor activities such as skydiving, snowboarding, hiking, mountaineering, climbing and scuba diving. He is also a licensed radio amateur

(KF5ONO) and has participated in several ARISS (Amateur Radio on the International Space Station) educational contacts.

The character of German ESA astronaut, Alex Vogel in the 2015 movie *The Martian*, is based on Alexander Gerst. They share appearance, first name and home town.

He was selected as an ESA astronaut candidate in May 2009, officially joined ESA in September 2009 and completed Astronaut Basic Training in November 2010.

The Blue Dot Mission

Mission Patches

The mission name and logo are inspired by an image of Earth taken by NASA's Voyager 1 spacecraft on February 14, 1990, from a distance of 6 billion km from our planet. American astronomer Carl Sagan asked NASA to turn the ship around so that it could take this photograph and described it as a "pale blue dot". In the patch, Earth is held and protected by two hands, emphasizing that it is the only "spaceship" that the human race has.

The Soyuz TMA-13M patch was designed by Luc van den Abeelen, based on the 'Fisherman' monument at Baikonur. In the mosaic, the cosmonaut is spreading his arms, as if in flight. Also, part of the design is a Soyuz rocket on display in the Russian town Samara where Soyuz rockets are being manufactured by TsSKB-Progress. It was originally offered to Soyuz TMA-05M commander Yuri Malenchenko in 2011 but Malenchenko went with a simpler design. It was then offered to Soyuz TMA-09M commander Max Suraev, who planned to use it. When he was replaced as commander by Fyodor Yurchikhin, the new commander had his own plans for a patch design so it was shelved again. Suraev, as commander of Soyuz TMA-13M finally approved it for use on that mission. Cepheus was the mission call sign and the Cepheus constellation and the Roscosmos logo were placed over the rocket. There is a discrete MZ initials in Earth's clouds to commemorate American friend of the crew and Soyuz patch collector Mike Zolotorow, who had died on December 10, 2013.

The Expedition 40 patch depicts the past, present, and future of human space exploration. The crew wrote the description, as follows: "The reliable and proven Soyuz, our ride to the International Space Station, is a part of the past, present, and future. The ISS is the culmination of an enormous effort by many countries partnering to produce a first-class orbiting laboratory, and its image represents the current state of space exploration. The ISS is immensely significant to us as our home away from home and our oasis in the sky. The commercial cargo vehicle is also part of the current human space exploration and is a link to the future. A blend of legacy and future technologies is being used to create the next spacecrafts which will carry humans from our planet to destinations beyond. The sun on Earth's horizon represents the new achievements and technologies that will come about due to our continued effort in space exploration."

Fig. 4.2. Blue Dot mission patch (ESA)

Fig. 4.3. Soyuz TMA-13M mission patch (www.spacefacts.de)

Fig. 4.4. Expedition 40 mission patch (ESA)

The Expedition 41 patch was designed by the crew and Blake Dumesnil. The crew wrote some text to go along with the design: "Portraying the road of human exploration into our vastly unknown universe, all elements of the Expedition 41 patch build from the foundation, our Earth, to the stars beyond our solar system. The focus of our 6 months expedition to the International Space Station is Earth and its inhabitants, as well as a scientific look out into our Universe. The distinguishing ISS solar arrays reach onward and serve as the central element, with the icon of an atom underneath representing the multitude of research on board that will bring new discoveries for the benefit of humanity. The Sun is rising over Earth's horizon, spreading its light along the road of human exploration. Equipped with the knowledge and inspiration gained from ISS, our successful multinational cooperation will lead human space exploration to the Moon, Mars, and ultimately, the stars. We are Expedition 41. Join us for the adventure."

Fig. 4.5. Expedition 41 mission patch (ESA)

Fig. 4.6. The Soyuz TMA-13M crew with Alexander Gerst on left (ESA)

Fig. 4.7. The Expedition 40 crew with Alexander Gerst third from right (NASA)

Fig. 4.8. The Expedition 37 crew with Alexander Gerst second from left (NASA)

Timeline

Expedition 40 Pre-flight

See Appendix 3 for a "Pre-Flight Interview with Alexander Gerst".

Expedition 40, Week Ending June 1, 2014

Soyuz TMA-13M launched from Baikonur at 3:57 p.m. EDT on May 28, 2014, carrying NASA's Reid Wiseman, RKA's Maksim Suraev and Alexander Gerst of ESA. After the, by now, routine 4-orbit chase, the Soyuz docked automatically to the Earth-facing port of the station's Rassvet Mini-Research Module-1. The hatches between the Soyuz and the ISS were opened at 11:52 p.m. EDT and the new crewmembers were welcomed by station commander Steve Swanson of NASA and Flight Engineers Oleg Artemyev and Alexander Skvortsov of Roscosmos, who had been on board the ISS since March 27. This was Suraev's second spaceflight and visit to the station. He was a flight engineer on Expeditions 21 and 22. For Wiseman and Gerst, this was their first spaceflight.

Alexander Gerst wrote in his ESA website blog,

I knew it would be impressive and loud – that much was clear. After all, I had seen the Space Shuttle launching from Cape Canaveral into the night sky years before. However, these impressions seem to fade away.

As a geophysicist and volcanologist who has worked on volcanoes for years, I thought to myself that seeing – or rather feeling – another rocket launch could not be more impressive than a volcano erupting at close quarters. But I had overlooked something.

If you are in a team, training for a mission to the International Space Station, you automatically become the backup for the crew flying six months earlier. For this reason, last November I flew to Baikonur with my two teammates, Max and Reid, in our capacity as backup crew. Two weeks before the launch of the main team, the Prime Crew, you have to go through the same quarantine and preparation programme, to be ready to become the substitute team in the event of an emergency.

You become very good friends with the Prime Crew through years of shared training. All this is intensified by the unique experiences shared shortly before flying into space, leaving this planet behind for half a year. The time spent in isolation in Baikonur is used to prepare mentally for the launch, to finalise your procedure books, to go through the final training exercises, to inspect the spacecraft and to attend to any last 'earthly' matters that one might have missed during two and a half years of intensive training.

The prime and backup crews part ways just two hours before the launch, after the prime crew has put on their space suits and you are driven to the launch pad of the Russian Soyuz rocket, which looks almost unreal in the glaring floodlights. This is the place from which Yuri Gagarin became the first human to leave Earth.

In this place, right in front of the launcher, I said goodbye to Misha, Rick and Koichi with a handshake, and watched my friends as they climbed into the spacecraft. At that moment, it became clear to me how different it would be if it were me who was going launch on this rocket with 26 million horse-power and fly into space within just eight minutes. The people on board are my friends – those with whom I had breakfast this morning and I have just helped into their space suits will have dinner on the Space Station tonight.

And this very fact is an aspect of the rocket launch that I will never forget. The memory of the plume of fire, the deafening rumble and the tremendous vibration of the launch are already a little faded in my mind. What remains is the memory of the stunning feeling that three people are sitting on that rocket, on their way to space.

*It was very quiet on the bus during our return from the launch site to Baikonur.
Such an event apparently makes us humans thoughtful – in a positive sense.
There are probably too many events to process in too short a time. What
caught me off guard was the welcome we received as we arrived at our
accommodation in Baikonur: "Welcome to the new Prime Crew." It was at
that moment that I realised – next time, I'll be the one on the rocket.*

With the ISS crew restored to full occupancy, with six crewmembers, the team
began their workday on May 30. Swanson reconfigured the Japanese Experiment
Module's scientific airlock in the Kibō module. He passed a new camera to the
outside of the station using the airlock's slide table. The ground team then commanded the Special Purpose Dexterous Manipulator install the camera near the
elbow joint of the Canadarm2 robotic arm.[1]

Wiseman, Suraev and Gerst familiarised themselves with their new home and
workplace. Wiseman used a Velocicalc tool to measure the airflow, temperature
and humidity throughout the U. S. segment of the station. He also replaced a tank
in the station's Water Recovery System. Gerst reconfigured the Saibo rack's Clean
Bench in the Kibō lab. He replaced a xenon lamp that did not switch on as expected
during a previous Resist Tubule experiment. Suraev unpacked cargo from the
Soyuz and participated in the Virtual study, which investigates how an astronaut's
spatial orientation is affected during a long-duration space mission. Skvortsov
packed rubbish and other unneeded items into the Progress 53. Artemyev upgraded
software for the Napor-mini RSA experiment, which uses an optical telescope and
radar to measure the Earth's environment. He also checked the ventilation in the
Russian segment of the station and downloaded micro-accelerometer data from
the Identification experiment to understand the dynamic loads of the ISS.

Expedition 40, Week Ending June 8, 2014

On June 2, Swanson and Gerst conducted eye checks for the continuing Ocular
Health study. Swanson worked on the Veggie experiment. He watered plant pil-
lows and checked the lighting on the botany science hardware. Wiseman worked
on the Advanced Colloid Experiment-Microscopy-2, which studies the behaviour
of microscopic liquids and gases separating from each other. Wiseman activated a
high-flow fluid transfer pump to transfer fluid to a wastewater bus. Gerst set up the
Resist Tubule experiment, checking lamps on the microscope. This observes gravity resistance in plants. Skvortsov stowed trash in the Progress 53 ship, while
Suraev off-loaded equipment and supplies from the Soyuz TMA-13M spacecraft.
Artemyev carried out maintenance in the Russian segment. He filled a water supply bag and maintained a network router.

---

[1] This operation, begun the previous week, made Dextre and Canadarm2 the first robot to repair
itself in space.

On June 4, Swanson set up a camera for the Resist Tubule study and worked with Wiseman using an ultrasound for eye scans. He then joined Gerst for blood pressure measurements and heart scans and installed an antenna cable on a U. S. spacesuit. Wiseman configured the Light Microscopy Module for the Advanced Colloids Experiment (ACE) fluids physics study. Gerst continued preparations for the upcoming Orbital-2 mission by assembling the Cygnus hardware command panel power and data cables in the Kibō. He also collected samples for the Microbiome study. Skvortsov and Artemyev reviewed their upcoming spacewalk timeline and watched an instructional DVD. Artemyev monitored urine transfers from the station's U.S. segment to a docked Progress spacecraft. Skvortsov worked on Russian hardware maintenance. Suraev collected blood and saliva samples for the Russian Chromosomes study. He then measured his arterial blood pressure.

On June 5, Suraev installed the Progress docking mechanism after final loading of trash was complete. He also transferred cargo from the Soyuz TMA-13M spacecraft. Gerst continued to test the Cygnus hardware command panel. The Cygnus cargo spacecraft was due to arrive at the end of June. Skvortsov and Artemyev collected tools and hardware for an upcoming Russian segment spacewalk. Station commander Swanson worked on servicing a U. S. spacesuit, checking for water and gas leaks to ensure the successful installation of a fan pump separator on the spacesuit. He also recharged batteries on spacewalk equipment and tools. Wiseman rotated a science freezer rack out of the wall inside the Destiny laboratory module so he could access and replace two faulty power control modules. He also switched temperature and humidity control from one common cabin air assembly to another inside Destiny.

Gerst filled out a questionnaire on headaches. Many astronauts have suffered from severe headaches and scientists are trying to discover why. He was one of thirty astronauts researching this area over a number of years.

Skvortsov and Artemyev closed the hatches of the Progress 53 craft on June 6. They prepared the Pirs docking compartment for the planned 19 June spacewalk. Swanson worked on the Resist Tubule and Veggie experiments. He later joined Gerst for Cygnus rendezvous training and then gathered trash to be later disposed of on Cygnus. Wiseman collected blood samples for stowage in a science freezer and then stowed the tools, jumper cables and the Remote Power Controller Modules he removed the previous day from a wall in the Destiny laboratory. He moved crew water containers then collected microbial surface samples from the interior panels of the Unity node. Suraev replaced dust filters and cleaned fan grids and removed equipment from a Soyuz vehicle descent module for return to Earth later.

Expedition 40, Week Ending June 15, 2014

The Progress 53 cargo craft undocked from the aft port of the Zvezda service module at 9:29 a.m. EDT on 9 June. Skvortsov photographed the departing cargo ship from the Russian segment. At 12:34 p.m. the deorbit burn ensured it was destroyed over the Pacific Ocean less than an hour later. This cleared the Zvezda docking port for the expected arrival in August of the fifth ESA Automated Transfer Vehicle, ATV-5, also called 'Georges Lemaitre'. Skvortsov and Artemyev continued to prepare for their spacewalk and Artemyev checked out Otilia experiment hardware for monitoring particle impacts on the station. He finished his day by downloading data for the Identification study, which measures the loads on the station during dynamic events such as Monday's Progress undocking. Swanson drew a blood sample from Gerst and installed a new test sample for the JAXA Resist Tubule experiment. Wiseman collected U. S. spacewalking tools and equipment to loan to Skvortsov and Artemyev for use during their spacewalk, including tethers and a pair of helmet cameras that can provide live, first-person views from the spacewalk. Gerst meanwhile participated in a periodic fitness evaluation while working on an exercise bike known as the Cycle Ergometer with Vibration Isolation and Stabilization System (CEVIS). Suraev removed a series of brackets in the Rassvet Mini-Research Module-1. Wiseman later swapped out a manifold bottle in the Combustion Integrated Rack. Before retiring, Wiseman and Swanson spoke with CBS Evening News' Scott Pelley from the Destiny laboratory.

On June 10, 2:40 p.m. EDT, there was report of smoke in the Zvezda Service Module. After isolating the Russian segment ventilation system, the smoke subsided and it was traced to a heater in the water reclamation unit used for food preparation. The crew powered down the unit and set up a fan and a filter. The smoke cleared and they took atmospheric readings throughout the station to assess any damage. Procedures were followed and there was no danger. The faulty part was removed and placed in the Progress 55 cargo ship.

Swanson replaced tubing inside the Water Recovery System and retrieved germinating seedlings from the Minus Eighty-Degree Laboratory Freezer for ISS (MELFI). Wiseman and Gerst measured each other's bodies as part of the Body Measures experiment. This investigates the changes to astronauts' body size during spaceflight. Gerst and Swanson carried out training exercises for the robotic capture of Orbital Sciences' Cygnus cargo ship and berthing at the Harmony node. Swanson harvested some red romaine lettuce plants growing in the Veggie facility, a low-cost plant growth chamber that uses a flat-panel light bank for plant growth and crew observation. Then he gathered tools and reviewed procedures for upcoming configuration of the Multi-user Droplet Combustion Apparatus Chamber Insert Assembly for the Flame Extinguishment Experiment-2. Wiseman deployed eight neutron monitors for the RaDI-N radiation detection study. Skvortsov and Artemyev configured their Russian Orlan spacesuits and reviewed the planned worksites for their upcoming EVA. Suraev carried out a leak check of the cooling loops in Zvezda and deployed new radiation dosimeters for the Matryoshka experiment.

On June 11, Swanson replaced components of the Multi-user Droplet Combustion Apparatus to configure it for the next round of data collection for the Flame Extinguishment Experiment-2 (FLEX-2). Results from the FLEX-2 Fuel Surrogate test points will lead to greater efficiency of liquid-fuel engines and minimize pollution. Gerst and Wiseman conducted the Cardio Ox study, to investigate the risks of cardiovascular disease related to long-duration spaceflight. Gerst also participated in the European Space Agency's Skin B experiment. This test evaluates skin aging mechanisms that are faster in space than on Earth. Swanson collected detailed imagery of Gerst's eyes using optical coherence tomography equipment. Afterward, Gerst checked the sizing of a new spacesuit planned to be used by Swanson on a future spacewalk. Wiseman changed out a recycle tank in the station's Water Recovery System and inserted 12 Ice Bricks into a MELFI. The Ice Bricks consisted of various fluids that are encapsulated in a high-density polyethylene capsule and are color coded for specific temperatures. Artemyev and Skvortsov resized the Russian Orlan spacesuits for their EVA. Artemyev also installed a spare Russian water heating and dispensing unit in the Zvezda service module following the previous day's smoke incident. The crew sent their best wishes to all the teams competing in the World Cup in Brazil, which took place from June 12 to July 13, 2014.[2]

Swanson removed a seedling sample chamber from the Cell Biology Experiment Facility Incubator Unit and prepared the sample for a close-up look with the Clean Bench's microscope as part of the JAXA Resist Tubule experiment. He later spoke with Pedro Echevarria of C-SPAN's "Washington Journal" show. Wiseman performed an ultrasound scan on his own thigh and calf muscles as part of the Sprint investigation. This experiment evaluates high-intensity, low-volume exercise training in minimizing the loss of muscle mass and bone density that occurs during long-term microgravity missions. He removed sensors and an armband monitor that tracked his activity for 36 hours for the Circadian Rhythms study and he logged his meals for the Pro K study as nutritionists monitor how dietary changes may affect spaceflight-related bone loss. Gerst checked out the new US spacesuit in the Quest airlock. This suit arrived on a SpaceX Dragon capsule and was the first to be delivered to the station after the retirement of the space shuttle. Skvortsov and Artemyev conducted the Russian BAR experiment to study methods of detecting a leak from one of the station's modules. Artemyev monitored dosimeter readings from the Matryoshka experiment and Suraev cleaned the fan screens in the Poisk and Rassvet modules.

Swanson began the June 13 with the Sprint investigation. Wiseman was the subject of the Cardio Ox study, which investigates the risks of cardiovascular disease on long-duration spaceflights. Gerst performed an ultrasound scan on Wiseman and measured his blood pressure. He also set up samples and configured hardware for a combustion experiment known as the Burning and Suppression of Solids (BASS) to investigate how materials burn in the absence of gravity. Swanson and Wiseman replaced a failed heat exchanger for the Common Cabin Air Assembly in the Tranquility node. In the Quest airlock, Gerst finished the checkout of the new spacesuit and he replaced the water as part of the maintenance routine for storage. Skvortsov and Artemyev borrowed tools and equipment from

---

[2] United States and Germany, both countries with crew on board, were scheduled to play each other on June 26 at Arena Pernambuco in Recife, Brazil.

their NASA crewmates for attaching to their spacesuits. Suraev and Skvortsov installed a docking mechanism for the Progress 55 cargo craft and audited the docking and internal transfer system tools and equipment.

On June 13, Gerst wrote in his ESA blog:

Exploring new horizons has never been easy. Throughout the hundreds of thousands of years of human history, people have been adventurers; some willingly and others forced by circumstances have taken great risks while seeking new prospects or places to live. Some paid for their journey of discovery with their lives, while others succeeded without incurring harm. Without these pioneers, our horizon would extend no farther than our legs are able to carry us in the course of a day.

Nowadays, we no longer set out in dugout canoes to paddle along the coastline until someone dares to build a larger ship and set sail beyond the horizon.

Or do we?

The International Space Station that I currently call home is precisely this kind of canoe, constructed in a fascinating act of international cooperation by over 100,000 people from many different nations. It flies 400 kilometres above our heads, outside Earth's atmosphere – but if we consider a globe for scale, it is only about the width of a finger away from our home planet. It will never fly further away – it was not designed to do so. Just once before, more than 40 years ago, we humans sent a handful of our compatriots a little further afield, to the next 'island'. But then our courage deserted us.

If we're lucky, this canoe will drift along 'our coast' for a few more years, serving as a research laboratory and vantage point from which we can peer towards the 'horizon'. Only in this way can we learn how to build a ship that will take us farther away. This joint project is of such immense importance to us that it remains sacrosanct, even in times of international tension. In fact, by letting us get a glimpse of all we could lose through conflict, it helps to bind us together.

The next 10 years will pass much faster than we expect. A few of us have started to take the knowledge we have acquired on board our 'canoe' and are using it to build a vessel that could someday carry us beyond the 'horizon', to the Moon and on to Mars. We might learn whether we have siblings out there in the Universe, as well as how we could spare Earth the fate of becoming a barren planet.

It will require a conscious decision, taken together as inhabitants of planet Earth, to take a ship where no one has gone before, beyond the 'horizon', to reach out and explore new worlds. The clock is ticking.

Engineers will be looking for new challenges when the International Space Station project comes to an end and, without further work for them, their valuable knowledge will be lost.

Venturing out toward new cosmic horizons would be the greatest adventure humanity has ever faced. Let's not miss the boat!

Expedition 40, Week Ending June 2, 2014

With a Russian spacewalk scheduled for June 19, Skvortsov and Artemyev set up their Russian Orlan spacesuits on June 16, with help from Wiseman. They installed borrowed NASA lights and cameras on the suits and checked the suits' telemetry and communications equipment. Later Wiseman logged what he ate for lunch then collected his blood and urine samples for stowage in a science freezer. He then started troubleshooting the lithium-ion battery chargers on the U. S. spacesuits. Gerst conducted two flame tests inside the Destiny laboratory module's Microgravity Science Glovebox as part of the Burning and Suppression of Solids combustion study. Swanson collected water samples from a potable water dispenser for inflight analysis after a morning exercise session. He then removed and replaced an air hose and liquid indicator from the Tranquility node's Waste and Hygiene Compartment (toilet) before lunchtime. He performed maintenance on the Advanced Colloids Experiment cleaning the work area after an experiment sample was broken during an earlier mixing session. After that work, Swanson was back inside the Tranquility node replacing a heat exchanger and purging a Sabatier accumulator. Max Suraev readied the Pirs docking compartment the upcoming spacewalk and assisted Skvortsov and Artemyev with Orlan suit emergency tasks in the unlikely event of a pressure leak inside Pirs.

As the spacewalk drew nearer, Artemyev and Skvortsov put on their Russian Orlan space suits on June 17, checked their systems and carried out a pressure check. Suraev assisted his colleagues and then worked on the Matryoshka experiment. Swanson charged batteries in tiny student-controlled satellites for an upcoming experiment run of the SPHERES-Slosh investigation. Then he filled and depressurized a flush water tank in the Harmony node's Waste and Hygiene Compartment. Wiseman opened up the Fluids Integrated Rack to prepare samples for the Advanced Colloids Experiment (ACE). He also set up the Light Microscopy Module inside the rack for the ACE study which observes the microscopic behaviour of liquids and gases separating from each other. Gerst measured his body mass using the Space Linear Acceleration Mass Measurement Device (SLAMMD). The device applies a known force to a crew member with the resulting acceleration used to calculate body mass. He later replaced cable arm ropes and applied the proper tension on the Advanced Resistive Exercise Device (ARED) and set up a video camera to record his cable troubleshooting efforts for the Burning and Suppression of Solids experiment.

Preparations continued on June 18 for the following day's Russian EVA. Suraev assisted his crewmates and then performed maintenance in the Pirs module, preparing cables, replacing filters and cleaning fan grilles in the docking compartment. Swanson and Wiseman worked on the SPHERES-Slosh experiment for students, executing test runs. Gerst recorded the Earth as part of the Education Payload Operations program. He then worked with the Fundamental and Applied Studies of Emulsion Stability (FASES) hardware. This is an ESA emulsion study in the Columbus module's Fluids Science Laboratory. The experiment could lead to more environmentally friendly products.

On June 19, Alexander Skvortsov and Oleg Artemyev spent 7 hours and 23 minutes on their Russian segment maintenance spacewalk. They fitted a new antenna on the exterior of the Zvezda service module. They moved Obstanovka experiment hardware which monitors the plasma and the magnetic field around the station. They swabbed the crew quarters window on Zvezda and they discarded an orbital debris experiment (MPAC) and a materials exposure experiment (SEED) attached to Zvezda's aft section. Finally, they relocated a payload cargo boom on Zvezda. They entered the Pirs module and closed the hatches at 5:33 p.m. EDT.

The following day, June 20, the crew was allowed late start after the spacewalk activities. Once at work, Swanson set up the Capillary Flow Experiment-2 in the Destiny laboratory module. He recorded his space journal and talked to ground controllers about the inventory management system. Wiseman loaded science freezer payload application software on an EXPRESS rack laptop computer. He then inventoried the Human Research Facility supply kits He took pictures of the contents. Gerst saved data from a portable, wearable electrocardiogram to a medical laptop computer. This was for the Biological Rhythms 48 study which investigates the crew's biological rhythms. He later demonstrated physics principles in space for the Educational Payload Operations suite of experiments. He videotaped soap bubbles in microgravity for a student programme. The cosmonauts carried out post-spacewalk tasks, such as deactivating the Pirs' spacewalk systems and installing ventilation ducting in the docking compartment, removing U. S. cameras and lights from the Orlan spacesuits, drying the water feed lines, discharging suit batteries and preparing the suits for stowage.

Expedition 40, Week Ending June 29, 2014

Swanson started the week on June 23 by conducting an ultrasound scan of his right thigh and calf. This was part of the Sprint experiment which records high-intensity, low-volume exercise training and how it can reduce the loss of astronaut muscle mass and bone density. Wiseman prepared test samples for the Advance Colloids Experiment (ACE). Gerst set up samples and configured hardware for the Burning and Suppression of Solids (BASS) experiment. Wiseman checked the torque settings on pistol grip tools. Artemyev stowed equipment after his and Skvortsov's spacewalk. Skvortsov installed digital TV cameras in Zvezda, while Suraev updated computer software.

For 2 years, fluctuations in the current being fed from two main bus switching units to the Zvezda's power converters have been recorded. On 24 June, Swanson and Skvortsov installed an oscilloscope to investigate these fluctuations and moved an electric current measurement clamp around the Zvezda to find the source of the problem. Wiseman stowed U. S. tools returned from the Russian spacewalk. Gerst prepared more samples in the Microgravity Science Glovebox for the Burning and Suppression of Solids (BASS) experiment. Gerst, Wiseman and Swanson spoke with reporters from Univision's "Contacto Deportivo" and ESPN's "SportsCenter" to discuss the 2014 World Cup games, specifically the upcoming Germany vs. United States match. Gerst and his American crewmates made a bet: If Germany lost the match Gerst promised to paint a U. S. flag on the side of his head. If Germany won, Wiseman and Swanson promised to shave their heads. Wiseman said "I believe we will win. It's two against one up here, so I think the U.S. chances are pretty good." Gerst replied "I hope we kick their butt a little bit, but I'm going to hope it's going to be at the final game, not at this game on Thursday."

Wiseman and Artemyev conducted another test run of the SPHERES robots in preparation for the Zero Robotics middle school tournament, where teams of students from across the United States would compete to write the best software to control the SPHERES. Swanson had an eye test and collected water samples from the internal thermal control system. Suraev inspected the interior surfaces of the Russian segment of the station. During the day, ground controllers moved the Flight Releasable Attachment Mechanism (FRAM) from External Stowage Platform-3 (ESP-3) to ESP-2 in advance of two U. S. spacewalks scheduled for August.

On June 25, Swanson continued the Ocular Health study while Wiseman conducted an ultrasound scan of the calf and thigh of his right leg for the Sprint experiment. In the Kibō laboratory Gerst worked on the Ice Crystal 2 experiment. This investigates the growth rates and stability of ice crystals in super-cooled water containing antifreeze glycoprotein. Then he transferred a multi-purpose experiment platform and a robotic arm known as the Small Fine Arm through the Kibō module's scientific airlock. Swanson, Wiseman and Gerst trained for the robotic capture of Orbital Sciences' Cygnus cargo craft after which Swanson and Wiseman talked with students and teachers gathered at the Reuben H. Fleet Science Center in San Diego, California. Artemyev continued the Virtual experiment. He also carried out maintenance on the life-support system in the Zvezda service module and finished his Crew Medical Officer (CMO) proficiency training. Skvortsov and Suraev used an eddy-current tester to examine the structure of the Russian segment to detect any flaws in the surfaces. Skvortsov also reconfigured the Zvezda work area after completing power diagnostics with Swanson over the previous days.

The Zvezda's thrusters were fired at 6:41 a.m. EDT for 1 minute and four seconds to place the station in the correct position for the launch of the Progress 56 cargo ship, scheduled for July 23. The reboost raised the station's altitude by 0.3 miles at apogee and 1.9 miles at perigee, leaving the station in an orbit of 262.2×254.1 statute miles.

On the day of the Germany vs United States World Cup match, June 26, Germany's Gerst examined American Swanson's eyes using optical coherence tomography equipment. Afterward Swanson and Gerst worked restored the flow of the Internal Thermal Control System's coolant to the Midbay Common Cabin Air Assembly's heat exchanger in the proper direction. Once that was completed, Gerst set up Kubik-3, a small, temperature-controlled incubator for self-contained, automatic microgravity experiments. Wiseman meanwhile spent much of his time working with a combustion experiment known as the Burning and Suppression of Solids (BASS). Then Swanson, Skvortsov and Artemyev reviewed emergency Soyuz descent procedures. Artemyev also conducted the Uragan Earth-observation experiment and the Bar experiment, the latter of which investigates station leak detection. Suraev, removed and replaced a GIVUS gyrometer in the Zvezda service module. He and Skvortsov also inspected the Russian segment of the station using an eddy-current testing device to detect any flaws in the surfaces. Gerst installed the Soret Facet experiment into Solution Crystallization Observation Facility in the Japanese Kibō module. This experiment uses the station's microgravity environment to get a better understanding of a thermal diffusion phenomenon known as the Soret effect. Wiseman carried out a photo survey the Canadarm2's Latching End Effector in front a hatch window in the Harmony node. During lunch the crew watched an uplink of the World Cup match, which was won by Germany.

Wiseman tweeted on June 27: "It was nice to wake up this morning and not worry about my hair. There isn't any." Americans, Steve Swanson and Reid Wiseman, each had their heads shaved by Germany's Alexander Gerst after their team, USA, was beaten 1-0 by Germany in the World Cup group stage. Gerst did not have to paint an American flag on his already shaved head as part of the bet. Swanson finished a series of eye tests. He performed an ultrasound scan of his eyes, collected electrocardiogram data and measured his blood pressure. He also participated in the Skin B experiment. Wiseman worked on the Burning and Suppression of Solids (BASS) experiment in Microgravity Science Glovebox. He carried out a photographic inspection of the second Latching End Effector (LEE) of the Canadarm2 robotic arm through a window in the station's cupola. This is the other end of the robot arm. The LEEs are regularly inspected. Gerst performed an interior corner flow test for the Capillary Flow Experiment in the Harmony module and then set up Kubik-3.

Fig. 4.9. Steve Swanson, Alexander Gerst and Reid Wiseman in the Unity node after Gerst shaved the heads of the other two. (NASA)

Expedition 40, Week Ending July 6, 2014

Swanson, with assistance throughout the day of 30 June from Gerst and Wiseman, refurbished the Carbon Dioxide Removal Assembly (CDRA) in the Tranquility module. First, they replaced the desiccant/sorbent bed. Then they looked for leaks and cleaned the filters. In the afternoon, Wiseman worked in the Waste and Hygiene Compartment, replacing a pre-treat tank and he recorded the fluid flow. Finally, Wiseman conducted a vision test using an optical coherence tomography laptop computer. Suraev, Wiseman and Gerst practiced their medical proficiency in the afternoon. They ran an emergency simulation. Skvortsov and Suraev worked on the Ekon study where they photographed the surface of the Earth to document the ecological effects of industrialization. Skvortsov set up the Splanh experiment for upcoming work with the gastroenterological study. During the morning, Suraev sent Obstanovka experiment data to ground controllers. Afterward, he pressurized the Elektron oxygen generator in the Zvezda module and activated it. In the afternoon, he set up the VIRU virtual training software to study for the Relaxation experiment using 3-D manuals rather than the actual science hardware. Artemyev had a hearing test and helped Skvortsov test communications equipment and headsets. He later tested backup communication equipment in Zvezda. During the afternoon, Artemyev stowed equipment for disposal in the Progress 55 cargo craft and updated the inventory management system. Gerst spoke to the 64th Lindau Meeting, at Lindau and Mainau Island, Germany where 37 Nobel laureates and 600 young scientists met to "to share their knowledge, establish new contacts and discuss relevant topics such as global health, the latest findings in cancer or Aids research, the challenges in immunology or future research approaches to medicine". Alex had attended the previous year's event as a guest and fulfilled his promise to send greetings from the station.

On July 1, Swanson emptied the waste water tank in the Urine Processing Assembly in the Tranquility laboratory and rotated a rack to replace a fluids control and pump assembly in the Water Recovery System. Wiseman conducted an eye test, measuring the intraocular pressure of his eyes and then recorded the station interior configuration as part of a six-monthly routine video exercise. He taped the U. S. segment's laboratories, airlock, nodes and the Quest airlock through the Zarya cargo module and ended at the Soyuz vehicles. This ensures ground controllers can identify safety issues such as emergency path blocks, ventilation blocks, fire hazards and limited access to safety equipment. Gerst inspected hatch seals, surfaces and handle mechanisms in the U. S. segment. He took photographs of hatches and along with Wiseman, collected water and air samples for the Environmental Health System checks. Skvortsov and Suraev again photographed Earth for the Ekon Earth observation study, documenting the ecological effects of industrialization on Russia and nearby countries. Artemyev collected his blood and saliva samples for the Chromatomass microbiology study, then helped Suraev work on the Kulonovskiy Kristall experiment that observes the dynamics of a charged particle system in the magnetic field of microgravity. Suraev participated in the Interactions study that observes the personal, cultural and national differences between crew members and how it affects crew performance. The entire crew then performed an emergency simulation exercise while Houston and Moscow mission control centres monitored their actions.

Swanson continued his plumbing tasks on July 2, when he replaced parts of the Waste and Hygiene Compartment. He also set up software and hardware for eye exams using an Optical Coherence Tomography (OCT) laptop computer. Wiseman and Gerst then carried out eye tests with the OCT equipment with help from mission control. Wiseman then worked during the day on a combustion experiment in the Microgravity Science Glovebox in the Destiny module. Gerst maintained science equipment in the Kibō laboratory. He took a microscope from the Multipurpose Small Payload Rack (MSPR) and stowed it. Skvortsov took blood and saliva samples for the Chromatomass microbiology study. Artemyev worked on the Splanh gastrointestinal study. This investigates how the digestive system adapts to microgravity. Suraev set up the Rassvet mini-research module for work on the Kulonovskiy Kristall physics experiment. He also inspected and photographed windows of the Zvezda service module and sent the images to mission control.

Swanson set up the Capillary Flow Experiment hardware on July 3. He then set up the Ultrasound 2 and performed an echocardiogram on Gerst's chest with help from doctors at mission control. It was back to the plumbing then as he removed and replaced pumps and pipes inside the Tranquility node's Waste and Hygiene Compartment. Wiseman continued to work on the Burning and Suppression of Solids (BASS) study with assistance from the primary investigator on the ground over video link. Gerst and Wiseman scanned each other's eyes with the Ultrasound 2 medical equipment. Then Gerst and Swanson conducted eye pressure and blood pressure checks as part of the Ocular Health study. Gerst then deployed an alternate station computer inside Tranquility for the Force Shoes study. The exercise experiment will measure the dynamic loads a crew member experiences on the Advanced Resistive Exercise Device. Skvortsov worked Zvezda's ventilation system and tested Earth to station video downlink systems. In the afternoon, he stored equipment in the Pirs docking compartment and updated the station's inventory management system. Artemyev used the Motocard crew mobility experiment while Suraev continued to photograph windows for inspection. Later, Suraev set up the Rassvet mini-research module for more work on the Kulonovskiy Kristall physics experiment.

Expedition 40, Week Ending July 13, 2014

The crew light-duty on Friday to celebrate Independence Day as they began the holiday weekend. They carried out housekeeping chores and exercise training. They also photographed Hurricane Arthur travelling U. S. East Coast. On Monday July 7, station commander Swanson got back to work and assisted Gerst with a blood test for the Human Research Facility. Gerst then processed his blood sample with a centrifuge and stored it in the Minus Eighty Degree Laboratory Freezer for ISS (MELFI), for future study on the ground. Swanson then set up a pair of acoustic dosimeters that he and Skvortsov would wear for 24 hour to measure the noise levels on the station. Later Swanson replaced a recycle tank in the Environmental Control and Life Support System's Water Recovery System, which recycles condensation and urine to potable water. He also deployed eight RaDI-N radiation detectors that had been initialized earlier by Suraev to help characterize the radiation environment aboard the station. Gerst tested the proximity equipment in the Kibō laboratory for communicating with the anticipated Orbital Sciences' Cygnus cargo craft, scheduled for arrival the following week. Meanwhile, Wiseman continued with the Burning and Suppression of Solids (BASS), to study how fires behave in weightlessness. He then joined Swanson and Gerst for a training session as they prepared for the robotic grapple and berthing of Cygnus. Skvortsov set up proximity communications equipment to be used for the docking of the ESA ATV-5 cargo ship.

On July 8 Swanson participated in an on-orbit hearing assessment and then cleaned the Cell Biology Experiment Facility and performed routine maintenance on the Combustion Integrated Rack and the Fluids Integrated Rack. Wiseman and Gerst started reviewed procedures and set up hardware for an upcoming SPHERES test. Wiseman also recharged the battery packs for the hoop-shaped hardware called the Resonant Inductive Near-field Generation System (RINGS). SPHERES and RINGS form an integrated formation flight experiment using electromagnetic fields and wireless power transfer between satellites. Afterwards, Gerst set up a laptop computer for the Multi-End-To-End Robotic Operations Network (METERON), which examines the operational and technical capability for the station crew to remotely control robots on Earth. Wiseman meanwhile performed two simultaneous runs with the Binary Colloidal Alloy Test Low Gravity Phase Kinetics-Critical Point experiment (BCAT-KP-1), to allow the principal investigator to assess the second sample's behaviour during the first sample's run. Suraev performed the VIRU experiment, which aims to increase the efficiency of training and experiment operations through the use of 3D "virtual" manuals aboard the station. Skvortsov stowed rubbish in the Progress 55 cargo ship for disposal. Swanson, Skvortsov and Artemyev conducted a fit check of their Kazbek seat liners inside the Soyuz TMA-12M. Skvortsov and Gerst then teamed up in the Zvezda service module to perform a checkout of the control panel and video systems for the following month's docking of the ATV-5. Gerst rounded out his day with the Capillary Flow Experiment. This examines how fluids flow across surfaces with complex geometries in microgravity. Swanson performed an inspection of the fibre optic cable jumpers in the vestibules of the Harmony node, which serves as the connecting port and passageway between the Destiny, Kibō and Columbus laboratories as well as for visiting cargo ships.

Gerst recorded a "Flying Classroom" educational video to demonstrate the foaming of water in microgravity on July 9. Later he worked on the Skin B experiment where he applied three dermatology tools on his arm. Wiseman and Gerst continued to prepare and review procedures for the SPHERES/RINGS experiment. Gerst helped Swanson to replace a rope on the Advanced Resistive Exercise Device (ARED). Skvortsov and Artemyev also recorded an educational video about the Otklik payload, which tracks the impacts of particles on the station's exterior using piezoelectric sensors. Swanson, Wiseman and Gerst talked with reporters from Time.com and the Time Warner Cable Network in Albany, New York. Suraev performed the Relaxation experiment, which studies chemical luminescent reactions in the Earth's atmosphere. Swanson checked out the backup Robotics Work Station Display and Control Panel inside the Destiny laboratory and then Gerst and Wiseman joined him to practice grapple procedures with Canadarm2 in preparation for the arrival of Cygnus. Wiseman executed four flame tests using the BASS equipment. Swanson used the eValuatIon and monitoring of microBiofiLms insidE the ISS (VIABLE). For this he touched and breathed on sample bags. The VIABLE study involves the evaluation of microbial biofilm development on space materials. In Zvezda, Skvortsov replaced cables and conducted a test of the Telerobotically Operated Rendezvous System (TORU), to prepare for the upcoming undocking of Progress 55. TORU can be used to manually guide a Progress if the onboard Kurs automated rendezvous and docking system fails. Suraev carried out the Kulonovskiy Kristall experiment as he gathered information about charged particles in microgravity. Artemyev collected air samples in Zvezda and Zarya.

July 10 saw the Orbital Sciences' Antares rocket and the Cygnus cargo craft rolled out to Launch Pad 0A at the Wallops Flight Facility in Virginia. Meanwhile, on the station, Wiseman installed the Centerline Berthing Camera System in the docking port of the Earth-facing side of the Harmony node to help mission control check alignment of Cygnus. Swanson meanwhile set up the Capillary Flow Experiment in the Harmony module to study the interior corner flow of a bubbly liquid in microgravity. Gerst began preparing the Gradient Heating Furnace for a new software load. He later donned a pair of shoes equipped with special sensors and exercised on the COLBERT treadmill to provide data for the Force Shoes experiment, which seeks to quantify the loads experienced by crew members during their daily 2.5 hours exercise sessions. Skvortsov continued loading trash and unneeded items into the Progress 55. Suraev spent part of his morning with the Kulonovskiy Kristall experiment.

Artemyev conducted a photo survey of the Vozdukh atmosphere purification system to help Russian specialists develop a plan for replacing one of its components. In Kibō, Gerst removed a sample cartridge from the Hicari experiment, which seeks verify a method for producing high-quality silicon-germanium crystals. Swanson swapped out cartridges in the Materials Science Laboratory's Solidification and Quench Furnace. This metallurgical research furnace "provides three heater zones to ensure accurate temperature profiles and maintain a sample's required temperature variations throughout the solidification process. This type of research in space allows scientists to isolate chemical and thermal properties of materials from the effects of gravity". Artemyev participated in the Interactions experiment. Afterwards, he downloaded data from dosimeters deployed throughout the Russian segment to track the crew's exposure to space radiation.

On July 11, Orbital Sciences Corporation announced the postponement of the launch of its Antares rocket and the Cygnus spacecraft until Sunday at 12:52 p.m. EDT. Swanson removed and replaced a filter in the Water Recovery System. Wiseman used the COLBERT treadmill while wearing shoes equipped with specials sensors to calibrate the equipment for the Force Shoes experiment. Gerst recorded a "Flying Classroom" video. He took a brief break from that activity to talk with students in his native country of Germany over the station's ham radio.

Gerst finished up his morning activities with an inspection of the portable emergency equipment to make sure that the fire extinguishers and other items remain in working condition. The engines of the Zvezda service module were fired for 1 minute and one second beginning at 10:53 a.m. Friday to raise the orbit of the station by 0.8 statute miles at apogee and 1.3 statute miles at perigee. This placed the station in an orbit of 260.9 × 256.6 statute miles. Artemyev meanwhile cleaned fan screens inside the Zarya module and recorded video for a Russian documentary about life aboard the station. Swanson, Gerst and Wiseman had a conference call with the teams on the ground to review the plan to unload cargo from Cygnus. Skvortsov continued loading Progress 55 with trash and unneeded items for disposal. Suraev meanwhile returned to the Kulonovskiy Kristall experiment to gather more information about charged particles in a weightless environment. Swanson, Skvortsov and Artemyev held a handover conference with their replacements, NASA astronaut Barry Wilmore and Russian cosmonauts Alexander Samokutyaev and Elena Serova.

On the July 12, 2014, the crew celebrated a milestone. The International Space Station was occupied continuously for 5000 days. Since the start of Expedition 1 on 2 November 2000, 214 people had visited, lived and worked on the station.[3]

Expedition 40, Week Ending July 20, 2014

Swanson worked on July 14 on the Advanced Colloids Experiment. Gerst and Wiseman set up flow sensors in the Kibō laboratory ventilation system. He then removed samples and swapped desiccant packs in a science freezer. He set up the Burning and Suppression of Solids (BASS) experiment and conducted two flame tests. Skvortsov packed discarded equipment and trash inside Progress 55 and updating the inventory management records. Artemyev worked on maintenance in the Russian segment and helped Wiseman set up hardware for the EarthKAM student experiment in the Zvezda service module. Suraev collected data from the Matryoshka-R Bubble radiation detection experiment. He took photographs of windows in the Poisk module.

Orbital Sciences' Cygnus commercial cargo vehicle was launched from Pad 0A, Wallops Field, Virginia, at 16:52 UTC on July 13, 2014, carrying almost 3000 pounds of supplies to the station. This was the second flight under the Commercial Resupply Services (CRS) contract with NASA and designated Orb-2. The spacecraft was named *Janice E. Voss*, after the NASA astronaut and Orbital Science employee.[4] In preparation for its arrival, Swanson and Gerst practiced using the

---

[3] At the time of writing, September 30, 2019, the station has been in orbit for 7619 days and continuously occupied for 6906 days, more than any other space station in history.

[4] Janice Voss flew in space five times on STS-57, STS-63, STS-83, STS-94 and STS-99. She died February 6, 2012, aged 55.

Robotics Onboard Trainer, or ROBoT, to practice grappling Cygnus with Canadarm2. Swanson replaced a filter in the Destiny laboratory's potable water dispenser. Gerst reviewed Research Facility medical supply kits and replaced anything out of date. Afterwards, he rerouted power and data cables from the Kibō module to the cupola for a hardware command panel for future communications with Cygnus during the rendezvous. He installed a mini-camera in the cupola to record the robotic arm in action. Skvortsov continued to pack Progress 55 with trash and unneeded items for disposal. Artemyev worked with the Vizir experiment. This helps targeting of Earth photography. He also tested the power supply circuits for the Napor-mini RSA experiment; a telescope and radar that monitors the Earth's surface. Suraev, used a trace contaminant analyser in the Zvezda service module to check for the presence of ammonia, carbon monoxide and formaldehyde.

At 6:36 a.m. EDT, July 16, Cygnus was grappled by the Canadarm2 operated by Swanson and assisted by Gerst. It was berthed to the Earth-facing side of the Harmony module and second stage capture was completed at 8:53 a.m. EDT. Swanson remembered his former colleague after whom the craft is named. He said "We now have a seventh crew member. Janice Voss is now part of Expedition 40. Janice devoted her life to space and accomplished many wonderful things at NASA and Orbital Sciences, including five shuttle missions. And today, Janice's legacy in space continues. Welcome aboard the ISS, Janice." Wiseman removed the Centerline Berthing Camera System. He pressurized the vestibule between Harmony and the Cygnus and carried out a leak-check. Then Swanson and Gerst opened the hatch to the vestibule and prepared to open the hatch. After the Cygnus capture and berthing, Swanson began upgrading Robonaut 2. Mobility upgrades, including a pair of legs, new helmet pieces new shoulder and elbow covers arrived over the preceding cargo delivery missions. Artemyev downloaded micro-accelerometer data from the Identification experiment and performed the VIRU experiment, which uses 3D virtual manuals instead of paper and laptop instructions. Suraev inspected the Zarya module while Skvortsov continued packing the Progress 55.

Wiseman and Gerst removed the controller panel assembly for the Common Berthing Mechanism at the Earth-facing port of the Harmony node where the Cygnus cargo craft was berthed. Swanson and Gerst then outfitted the vestibule leading to Cygnus' hatch and equalised the cabin pressure between the station and the cargo craft. Swanson opened the hatch to Cygnus at 5:02 a.m. and inspected the interior of the vehicle to clear the way for Wiseman to begin unloading nearly 3300 pounds of supplies. Gerst set up the NanoRacks module which arrived on the cargo craft. NanoRacks provides a low-cost avenue for microgravity research through the use of standardised "plug and play" modules that fit neatly into a set of research racks aboard the space station. The crew also unpacked fresh fruit, a treat for the crew, usually loaded on each cargo spacecraft. Wiseman tweeted "Just sunk my thumb into the skin of a fresh orange and the smell completely over-whelmed my senses. Didn't know how much I missed real fruit." Wiseman also swapped out a lens on the EarthKAM camera mounted in one of the station's windows. Earth Knowledge Acquired by Middle School Students allows students to program a digital camera aboard the station to photograph a variety of geo-graphical targets for study in the classroom. The project was initiated by Sally Ride, America's first woman in space, in 1995 and originally called KidSat. Gerst continued to evaluate the Force Shoes, shoes equipped with sensors to measure the loads, while working out on the Advanced Resistive Exercise Device (ARED). This experiment was an engineering evaluation to see if these shoes can provide ARED load data that has been unavailable since 2011. He also disconnected air ducts from a rack in the Kibō module in order to remove a failed pump later in the week. Swanson performed more upgrades on Robonaut 2 and then talked with Jennifer Broome of KDVR-TV in Denver. Swanson grew up in the nearby city of Steamboat Springs, Colorado. Skvortsov continued preparing a Progress cargo ship for its departure and Suraev joined Skvortsov later in the afternoon to install a docking mechanism for Progress 55. Suraev also performed a monthly inspec-tion of the structural elements and cables inside the Zvezda service module. Artemyev conducted an audit of the lights in the Russian segment of the station.

July 18 was Swanson's third day working on the Robonaut 2. He installed new processors and removed and replaced fans, a power distribution board and other components inside Robonaut's torso. He also reviewed the plan for the testing of new lightweight, commercially available clothing designed to resist odours. Astronauts station clothing is not washed and they re-wear their clothing for more than one day. Swanson, Skvortsov and Artemyev were preparing to test the new clothing. After assisting Swanson with some of the Robonaut upgrades, Wiseman wore the Force Shoes to see if these shoes can provide ARED load data. He also deployed a new Tissue Equivalent Proportional Counter in the Zvezda service module to measure the radiation environment inside the station. Wiseman and Swanson participated in a live conversation about the future of space exploration with actor Morgan Freeman at NASA's Jet Propulsion Laboratory in Pasadena, California. Gerst spent most of his day replacing a thermal control system pump package in the Kibō module. He pulled a refrigerator-sized Environmental Control and Life Support System rack from the wall to remove and replace the failed pump package. Afterward, he moved the rack back into place and reconnected the air ducts. Skvortsov completed loading the Progress 55 craft with trash and unneeded items and closed its hatch in preparation for its undocking. Artemyev spent much of his day inspecting and photographing the windows inside Pirs and the Poisk Mini-Research Module-2. Suraev, continued a monthly inspection of the structural elements and cables inside Zvezda before moving on to remove a protective curtain in one of the crew quarters for the Matryoshka radiation-monitoring experiment.

Expedition 40, Week Ending July 27, 2014

The Progress 55 cargo craft undocked from the Pirs docking compartment at 5:44 p.m. EDT on July 21. After undocking, it moved to a safe distance from the station for 10 days of engineering tests. Pirs was now clear for the arrival of Progress 56, scheduled to launch two days later. In preparation for the arrival of Progress 56, Skvortsov and Suraev practiced using the Telerobotically Operated Rendezvous System (TORU), a manual backup for the Kurs automated rendezvous system. Swanson and Gerst spent had medical tests to evaluate and investigate the effects of long-duration spaceflight on astronauts. Gerst drew blood from Swanson for the Human Research Facility study, and Swanson processed the sample and stored it in the MELFI. They then tested each other's eye pressure as part of the Ocular Health study.

Swanson and Artemyev installed ultrasonic measuring equipment for a joint U. S.-Russian study. Meanwhile, Gerst constructed the Aquatic Habitat for the Zebrafish Muscle study. This studies the molecular changes that cause muscles to atrophy in space. Wiseman set up the Combustion Integrated Rack for ground-controlled tests. Wiseman later conducted another session with another combustion experiment, the Burning and Suppression of Solids (BASS). Wiseman and Swanson talked with NASA officials and Apollo astronauts gathered at the Kennedy Space Center in Florida for the renaming of the centre's Operations and Checkout Building in honor of Neil Armstrong. Present were NASA Administrator Charles Bolden, Kennedy Center Director Robert Cabana, and Mike Collins, Buzz Aldrin of Apollo 11 and astronaut Jim Lovell of Apollo 8 and Apollo 13. Lovell was the Apollo 11 back-up commander. Swanson and Wiseman talked about how Apollo 11 inspired them to become astronauts.

On July 22, the thrusters of the Zvezda service module were fired for 32 seconds starting at 6:57 a.m. EDT to slightly lower the orbit of the station to avoid debris from a Russian Breeze-M upper stage used to launch a Russian satellite in December 2011. The manoeuvre lowered the station's orbit by 1.1 statute miles at apogee and 0.1 of a statute mile at perigee and left the station in an orbit of 258.8 × 256.9 statute miles.

Swanson conducted an eye exam for the Ocular Health study and took an ultrasound scan of his right thigh and calf for the Sprint experiment. He then performed an optical coherence tomography exam on Gerst's eyes. Wiseman removed sensors and an armband monitor that had tracked his body's core temperature over a 36 hours period for the circadian rhythms study. Swanson and Wiseman talked to flight controllers about the timeline for a spacewalk scheduled for August. Suraev inspected windows in Pirs and the Zvezda service module. He also inspected and cleaned several laptop computers and recharged the satellite phone in the Soyuz TMA-13 spacecraft docked to the Rassvet Mini-Research Module-1. Skvortsov recharged the satellite phone in the Soyuz TMA-12M before moving onto the

Virtual experiment, a study of the vestibular system's adjustment to weightlessness. Skvortsov also downloaded data from the Identification experiment, which tracks the dynamic loads on the station during events such as dockings or reboosts. Artemyev replaced flow meter components in the Vozdukh atmosphere purification system. He later joined Skvortsov for some maintenance work on a communications panel. Artemyev rounded out his workday setting up a thermostat for the Kaskad cell cultivation experiment. He also participated in the Vzaimodeystviye (Interactions) experiment.

The Progress 56 launched from Baikonur at 3:44 a.m., Baikonur time, on July 23, carrying 3 tons of cargo. Four orbits and less than 6 hours later, it automatically docked to the International Space Station's Pirs docking compartment at 11:31 p.m. EDT. The craft was loaded with 800 kg of propellant, 21 kg of oxygen, 26 kg of air, 420 kg of water and 1319 kg of spare parts, experiment and supplies.

Swanson and Gerst took more eye tests and Swanson temporarily removed the Multi-user Droplet Combustion Apparatus from the Combustion Integrated Rack's combustion chamber to replace some igniter tips. He helped Wiseman with the ongoing Sprint exercise study. Wiseman photographed test samples for the Binary Colloidal Alloy Test. Gerst continued the Burning and Suppression of Solids experiment. Gerst also applied more dermatology tools to his arm as part of the Skin B experiment. Suraev and Artemyev examined the veins in their lower legs to provide data on the body's adaption to long-duration spaceflight.

After the arrival of the Progress 56 resupply spacecraft, the crew had a late start on July 24. Swanson cleaned the Clean Bench, or the Saibo experiment rack's glovebox in the Kibō laboratory. Wiseman recharged spacesuit batteries, then joined Swanson to talk with U.S. House Committee on Science, Space, and Technology Chairman Representative Lamar Smith and committee members. They discussed "the importance of space station research and technology, getting students interested in STEM fields and the station's role in setting the path for America's next giant leap to send humans to Mars". Gerst prepared for the launch of CubeSat nanosatellites by installing the NanoRacks CubeSat Deployer hardware onto the Multipurpose Experiment Platform in the Kibō module's scientific airlock. The recently arrived Cygnus cargo ship delivered 16 deployers containing 32 cubesats, including 28 Dove nanosatellites built and operated by Planet Labs, Inc., for a humanitarian Earth-imaging program. Artemyev worked on the Kaskad cell cultivation experiment and maintained the Russian segment toilet. Artemyev, Skvortsov and Swanson wore and evaluated the garments designed to resist odors.

Gerst continued to configure the Aquatic Habitat in Kibō laboratory for the Zebrafish Muscle study on July 25. Wiseman checked out the Multi-Gas Monitor by testing the ammonia and temperature sensors. Skvortsov continued unloading some of the 2.8 tons of cargo from the Progress 56 resupply spacecraft. Suraev used the Calcium experiment to measure loss of bone density due to microgravity. Artemyev began the day with a Crew Medical Officer proficiency training session.

He later conducted an audit of the spacewalk tethers aboard the station. He then downloaded micro-accelerometer from the Identification experiment, which measures the loads on the station during dynamic events and he monitored readings from the Matryoshka radiation-detection study. Artemyev rounded out his workday recharging a battery for the Cascade cell-cultivation experiment.

Expedition 40, Week Ending August 3, 2014

July 28 started with Swanson photographing Wiseman and Gerst who were taking blood samples. He then loaded software for the Human Research Facility-2 rack. Wiseman and Swanson carried out spacesuit maintenance, scrubbing cooling loops and collecting a water sample for analysis. Swanson removed alignment guides in the Combustion Integrated Rack. Afterward, he installed a plant experiment unit inside the Cell Biology Experiment Facility for the Resist Tubule botany experiment. Gerst conducted two flame tests reducing the oxygen partial pressure in the Destiny laboratory's Microgravity Science Glovebox (MSG) to create a stable blue flame with a long burn time. Scientists on the ground observed the work with cameras downlinking the video from inside the MSG. He also worked on the Kobairo experiment rack, measuring the insulation resistance of the racks' Gradient Heating Furnace which is used to investigate crystal growth on semiconductors. Skvortsov started his day with photography work for the Aseptic experiment which studies ways to sterilize space hardware. He then sampled and sterilized surfaces in the station's Russian segment for the microbiology experiment. Skvortsov later worked on Progress 56 cargo transfers and updated the inventory management system. Artemyev set up and worked throughout the day with the Kaskad experiment. Suraev started his day with Progress 56 cargo transfers. Next, he moved on to an experiment that explores using 3-D interactive manuals to train for experiments aboard the space station. Suraev then assisted Artemyev photographing his work for the Kaskad experiment. Finally, he participated in the Kaltsiy experiment, or Calcium, that observes bone demineralization in long-term space station crew members.

ESA's fifth and final Automated Transfer Vehicle (ATV-5) launched on July 29 on an Ariane 5 rocket from Kourou, French Guiana, at 7:47 p.m. EDT. It carried 7 tons of science, food, fuel and supplies. This spacecraft is named after the Belgian astronomer, Georges Lemaitre, who first proposed the expansion of the universe and applied Albert Einstein's theory of general relativity to cosmology.

While Swanson, Wiseman and Gerst transferred cargo from the Cygnus on July 30, mission control implemented an artificial 50 seconds communication delay to simulate a crew traveling beyond low-Earth orbit. Later Swanson measured the conductivity of water samples taken from the spacesuit cooling loops and collected tools in preparation for upcoming Russian and U. S. spacewalks. Wiseman loaded software for the Human Research Facility Rack-1 inside the ESA Columbus laboratory. Then he drained a Tranquility node recycle tank and checked the device for leaks. Gerst worked on the Burning and Suppression of Solids (BASS-II) combustion experiment. Suraev continued unloading the Progress 56 resupply ship and updated the station's inventory management system. Later he installed overlay plates on Zarya cargo module interior panels and treated them with disinfectant. Skvortsov worked throughout the day on maintenance tasks. He replaced a valve and checked for hydrogen leaks on the Elektron oxygen generator. He prepared to replace fans in the Russian modules with low noise units and he worked with the Motokard investigation which observes how a crew member adapts to moving around in microgravity during a long duration mission. Artemyev Suraev for the Zarya installation work.

Gerst took part in a demonstration experiment for the European Space Agency's Flying Classroom program. He videotaped the educational investigation of basic fluid physics for the Marangoni Convection experiment. Swanson transferred cargo from the Cygnus cargo craft to the Harmony node. He also participated in a study investigating the possible long-term risk of atherosclerosis in astronauts. The experiment named Cardio Ox analyses a crew member's blood and urine samples including Ultrasound scans of carotid and brachial arteries before, during and after a mission. Suraev was back at work transferring cargo from the Progress 56, updating the inventory management system as he went. Skvortsov continued the fan replacement work he started earlier in the week. He swapped the old fans with new low noise units throughout the space station's Russian modules. Artemyev worked in the Russian segment on various science and maintenance activities.

On August 1, after ten days of engineering tests, the Progress 55 was deorbited. Swanson fitted the U. S. spacesuits for himself, Wiseman and Gerst for upcoming spacewalks. Wiseman worked on the Microgravity Science Glovebox conducting more flame studies for the BASS-II experiment. Gerst transferred cargo from the Cygnus. He did some temporary stowage work inside the Unity node. He then entered the Z1 truss structure to collect equipment for the upcoming spacewalk and inspect hatch seals. Skvortsov configured communications equipment inside the Rassvet mini-research module throughout the day. Afterward, he went inside the Poisk mini-research module to replace old fans with new low noise units and conducted acoustic measurements. Artemyev did some ham radio work during the morning and later he did some sample mixing for the Kaskad microbiology experiment. Suraev continued more cargo transfers from the Progress 56.

Expedition 40, Week Ending August 10, 2014

On August 4, Swanson prepared the Combustion Integrated Rack for another ground-commanded session of the Flame Extinguishment Experiment-2 (FLEX-2). He then swapped out a recycle tank in the Water Recovery System (WRS) of the station's Environment Control and Life Support System. Later, Gerst, Swanson and Wiseman had a "wet helmet" training session as the three crewmates familiar-ized themselves with the absorption pad and snorkel outfitted in the helmets of the U. S. spacesuits. This precautionary safety equipment was added as a result of the water incursion experienced by Expedition 36 astronaut Luca Parmitano during his spacewalk in July 2013. Skvortsov and Artemyev spent the morning reviewing procedures for their own scheduled spacewalk. Suraev transferred cargo from the Progress 56. After lunch, Gerst and Skvortsov reviewed rendezvous contingency scenarios for the arrival of the ESA ATV-5. Swanson and Wiseman spent much of their afternoon configuring tools for their spacewalk. Afterward, Swanson received a set of dosimeters from Suraev and deployed them around the station to help categorize the radiation environment.

On August 5, Swanson had the glamourous task of transferring urine from a wastewater tank to a recycle tank for the Urine Processor Assembly. They then set up the Combustion Integrated Rack for more ground-commanded research and loaded trash and unneeded items into the Cygnus cargo craft. As the Burning and Suppression of Solids-II experiment was completed, Gerst removed its hardware from the Microgravity Science Glovebox in the Destiny laboratory. He then down-loaded data from monitors that he had worn to track his body's core temperature over a 36 hours period for the Circadian Rhythms study. Skvortsov and Artemyev gathered tools and hardware for their spacewalk. Suraev replaced hoses for the toilet in the Zvezda service module. Wiseman installed hardware for the Capillary Channel Flow experiment inside the newly vacated Microgravity Science Glovebox. Gerst and Skvortsov trained using docking simulations to prepare for the arrival of the ATV-5.

ISS program managers decided to postpone the U. S. spacewalks planned for August 21 and 29 until new long-life batteries were delivered on a SpaceX com-mercial resupply services flight. This was due to a potential issue with a battery fuse in the spacesuits.

On August 6 Swanson carried out routine maintenance on the Waste and Hygiene Compartment, and then he participated in more ocular health exams. He packed the spacewalk tools and removed batteries from the spacesuits. He pre-pared a different spacesuit for the replacement of fan pump. Gerst unpacked sup-plies from the Progress 56 resupply ship and also packed trash into the Cygnus cargo craft. Skvortsov and Artemyev prepared the replaceable components of the Orlan spacesuits to be worn during their spacewalk. They also configured the Pirs docking compartment airlock and collected EVA tools. Suraev refilled the tank for the Elektron oxygen-generating system and performing routine maintenance on the life support system in the Zvezda service module. Wiseman talked with Maryland Public Radio station WYPR and Maryland Public Television. He is from Baltimore, Maryland. Gerst recorded another educational video as part of the Story Time project. Then he maintained the Waste and Hygiene Compartment and replaced audio equipment in the Harmony module.

On August 7, Swanson started a program of upgrades to the Robonaut 2 in the Destiny module, replacing processors, circuit boards, cables and fans. This was to prepare for the future installation of mechanical legs on the robot. Wiseman mixed test samples for the Canadian Binary Colloidal Alloy Test (BCAT-C1). He later helped Gerst with the stowage of trash and unneeded items in the Cygnus. Suraev reinstalled communication system amplifiers in the Zvezda service module and helped Skvortsov and Artemyev with hand ergometry assessments. Then Skvortsov and Gerst had simulator training as they prepared to watch over the automated rendezvous and docking of the ATV-5. Gerst talked with ABC News' Gina Sunseri and Bill Harwood of CBS News from the Columbus laboratory. Then Gerst remotely controlled the Eurobot at the European Space Research and Technology Centre (ESTEC) in the Netherlands as part of a technology demonstration known as the Multi-End-To-End Robotic Operations Network (METERON). Suraev tested the secondary power supply for the Napor-mini RSA experiment. He also maintained the SOZh life-support system in Zvezda.

Expedition 40, Week Ending August 17, 2014

On August 11, Skvortsov and Artemyev sized their Orlan spacesuits and checked for leaks in advance of their EVA. Swanson, Wiseman and Gerst used the Space Linear Acceleration Mass Measurement Device (SLAMMD) to calculate their bodies' mass. Gerst reviewed the ATV-5 cargo transfer list. Wiseman continued troubleshooting a lithium-ion battery charger in a U. S. spacesuit. He then had an interview with WJLA-TV in Washington, D. C. He spent the rest of his afternoon working on science equipment inside the Combustion Integrated Rack for mainte-nance. He replaced a fuel reservoir that contains the liquid fuel necessary for the rack's droplet combustion experiments. Swanson continued the mobility upgrades for the Robonaut 2 humanoid robot. Suraev set up dosimeters for the Matryoshka radiation exposure experiment. He then made preparations inside Zvezda ahead of Tuesday morning's arrival of the ATV-5. Skvortsov, Artemyev and Suraev partici-pated in a study of their cardiovascular system, exercising on the Velo-ergometer, or exercise bicycle, under a graded physical load.

After a two-week chase, the Automated Transfer Vehicle (ATV-5) 'Georges Lemaître' arrived at the ISS on August 8 and docked automatically to the Zvezda service module's aft port on August 12 at 9:30 a.m. EDT. At first, the ATV posi-tioned itself less than four miles beneath the station to test new rendezvous sensors and laser systems before taking up a position above and behind the station for the final phase of its rendezvous.

With the ATV docked to the aft port of the Zvezda module, there were five spaceships docked at the station, the other four being two Soyuz, a Progress and a Cygnus. Gerst and Skvortsov entered the ATV on 13 August and prepared for unpacking by sanitizing the interior. Wiseman and Swanson measured their bodies for the Body Measures study. Wiseman installed blood pressure and electrocardio-gram equipment as part of the Sprint VO2 study. Gerst looked over the installation plan for new experiment payloads in the Columbus laboratory's European Drawer Rack. Later he spoke over ham radio with pupils in Ontario, Canada. Artemyev mixed biological samples in a bioreactor for the Caskad microbiology experiment. He also cleaned gas-liquid heat exchanger fan screens. Later he took samples of the air in the ATV to check carbon dioxide levels. Suraev worked on life support system maintenance and replaced a water and distribution heating unit in Zvezda.

Gerst opened the ATV-5's hatches permanently on August 14. He joined Swanson and they worked set up the Electromagnetic Levitator (EML) in the Columbus laboratory. They installed it in the European Drawer Rack. The EML allows the crew to investigate the properties of high-tech alloys and semiconduc-tor materials in a melted state. Swanson and Wiseman depressurized the vestibule between Cygnus and the Harmony and installed new equipment required when Cygnus unberths later. Skvortsov and Artemyev collected tools and prepared the Orlan spacesuits' communications and telemetry equipment for their upcoming EVA. They planned their work paths and areas. Suraev aided them in setting up their telemetry equipment. Suraev prepared the Pirs module for the excursion and studied bone calcium loss in microgravity.

The Cygnus cargo ship was released by the Canadarm2 robotic arm at 6:40 a.m. EDT on August 15. Gerst and Wiseman controlled the robot arm. Gerst then replaced a pre-treat tank and hose in the Tranquility node's Waste and Hygiene compartment. At the same time, Swanson and Wiseman participated in the SPHERES-Zero-Robotics student competition. The Russian cosmonauts had the day off to do some housekeeping and to continue to prepare for the Russian EVA.

Expedition 40, Week Ending August 25, 2014

Alexander Skvortsov and Oleg Artemyev began their spacewalk at 10:02 a.m. on August 18. As a first task, Artemyev manually launched Chasqui 1, a Peruvian nanosatellite with two image sensors to record the Earth's surface. They installed the EXPOSE-R2 experiment package on the outside of the Zvezda hull. This con-tained two astrobiology studies to investigate biomaterials and extremophiles, organisms that can survive the rigours of exposure to space. They installed a hand-rail clamp holder for the Automatic Phased Array antenna on the Zvezda. They installed the Plume Impingement and Deposit Monitoring unit on the Poisk Mini Research Module-2. They retrieved several science packages designed to expose a variety of materials to the harsh environment of space. On the Poisk module, they replaced cassette on the SKK experiment, a Russian materials study. They also collected a panel of sample materials from the Vinoslivost payload and a can-ister from the Biorisk experiment. They took samples of residue from one of the Zvezda portholes. The EVA lasted 5 hours and 11 minutes and was the 181st in support of space station assembly and maintenance.

Swanson remained in the Poisk modules, as is standard practice, during the EVA. Inside, he photographed the starboard solar arrays for further inspection by ground controllers. Wiseman set up new test samples in the Binary Colloidal Alloy Test.

Cygnus burned up on re-entry at 8:34 a.m. Sunday and the crew photographed it as it broke up over the Pacific Ocean at 9:22 a.m.

Gerst restowed items stowed in the Quest airlock in anticipation of the arrival a new Nitrogen Oxygen Recharge System (NORS) at a later date. Wiseman meanwhile checked the portable emergency provisions such as fire extinguishers and portable breathing apparatuses.

The crew had a late start on August 19 to make up for busy spacewalk activity the previous day. Skvortsov and Artemyev removed the U. S. helmet cameras and lights from their Russian Orlan spacesuits and handed them back to Swanson. They spent time drying out their spacesuits. Skvortsov later packed up some of the science samples retrieved during Monday's spacewalk for return to Earth. Artemyev meanwhile stowed medical equipment and began the discharge of an Orlan spacesuit battery to condition it for storage. Suraev performed routine maintenance of the life-support system in the Zvezda service module. Swanson and Wiseman assembled the external television camera group hardware to be installed on the station's truss segment during a spacewalk that was expected to take place in October. Wiseman spoke with WJZ-TV in his hometown of Baltimore, Maryland. In the Kibō module, Gerst operated the airlock and slide table in the Kibō's Multi-Purpose Experiment Platform to help ground operators launch NanoRacks cubesat miniature satellites. The CubeSat deployer mechanism was passed outside the station and grappled by the Japanese robotic arm. The first pair in this second batch of Planet Labs Dove satellites were ejected into orbit at 2:25 p.m. Wiseman was in the cupola to photograph Earth-imaging satellites as they floated away.

Swanson, Skvortsov and Artemyev, wearing their Sokol suits on August 20, carried out a fit check of the Kazbek seat liners inside their Soyuz TMA-12M in preparation for their return to Earth in September. Swanson moved on to the Sprint experiment. Wiseman mixed samples for the Binary Colloidal Alloy Test (BCAT-C1) experiment and then participated in the Body Measures experiment. Suraev reopened the hatch of the Progress 56 ship for the first time after the recent spacewalk. Swanson, Wiseman and Gerst unloaded more supplies from the ATV-5. Gerst helped and Swanson to move the Maintenance Work Area in the Columbus laboratory to a new location and then Gerst set up the Aquatic Habitat for the Zebrafish Muscle study. Skvortsov and Artemyev packed away their tools and equipment after their EVA on Monday. Then Suraev installed a computer to control the EXPOSE-R experiment package they had installed on the exterior of the Zvezda module. The crew held a conference call about server-client upgrades with Houston mission control. Gerst replaced a hard-drive and Wiseman networked computers so that ground controllers could install new software.

Wiseman and Suraev started the process of installing new hard drives in station laptops so that the ground controllers update the software remotely. Afterward, Wiseman and Swanson unloaded more supplies from the ATV-5. Swanson carried out routine maintenance on Tranquility's Waste and Hygiene Compartment and moved Contingency Water Containers in anticipation of work on the water recycling system. Gerst spent the day installing the Electro-Magnetic Levitator (EML) into the European Drawer Rack. The EML is a furnace that melts free-floating alloy samples and then allows them to solidify in an electromagnetic field chamber. Later, Suraev, Skvortsov and Artemyev held an interview with Moscow TV in the Kibō. They talked about Russian Mission Control. Afterward, Suraev continued the Uragan Earth-observation experiment. Swanson and Wiseman unloaded more cargo from the ATV-5 cargo. Later Wiseman loaded the TMA-12M spacecraft with U. S. items for its return to Earth.

The launch of the fourth and fifth pairs of Planet Labs Dove satellites were unsuccessful over Wednesday and Thursday. Ground controllers at Marshall Space Flight Center in Huntsville, Alabama and flight controllers at the Tsukuba Space Center, Japan tried to find answers. The next pair of satellites launched successfully at 9:37 a.m. to join Flock 1B. Further attempts to launch the stalled satellites were unsuccessful.

Wiseman set up computers and network equipment to prepare for the upgrade of software on the Station Support Computers on August 22. Gerst continued the Body Measures study and then carried out a periodic fitness evaluation by using the Cycle Ergometer with Vibration Isolation and Stabilization (CEVIS). Wiseman and Gerst took blood pressure and electrocardiogram readings to investigate the crew's cardiovascular and musculoskeletal health. Gerst answered social media enquiries in the ESA Columbus module and participated in a televised recording. He was asked what advice he could offer the world thanks to his orbital perspective, Gerst replied, "I don't think I should give advice to anybody, because who would listen to a guy who flew into space just because he flew into space? My approach to this is actually let the people on Earth see the Earth through my eyes. And that's why I tweet so many pictures." He later removed the Facility for Absorption and Surface Tension (FASTER) from the European Drawer Rack in the Columbus module. Swanson moved the Payload Data Handling unit in Kibō and Wiseman finished the computer and network upgrades. Skvortsov and Artemyev studied the veins in their lower legs and Suraev transferred water from the Progress 56 and updated one of the Station Support Computers after it received its new software load.

Swanson and Wiseman replaced a fan pump separator on a U. S. spacesuit on August 25. Swanson then joined Skvortsov and Artemyev to adjust their Kentaur suits. These are anti-gravity suits to prevent blood from pooling in the legs. Artemyev operated the Virtual study to see how the crew visually adapt to the potentially disorienting effects of spatial position in weightlessness. Suraev backed up a Russian computer and carried out a software update. He took samples from the interior of the Zarya cargo module and joined his crewmates in making a recording for Russian television.

Surprising the crew and ground controllers alike, the pair of nanosatellites that had failed to deploy in the previous week spontaneously launched on Saturday. Ground controllers powered down the NanoRacks cubesat Deployer on the exterior of the Kibō laboratory module. They concluded there were no safety issues for the crew.

Expedition 40, Week Ending August 31, 2014

Swanson set up the Robonaut 2 and its tools and hardware for ground commanded operations on August 26, before moving to the Columbus laboratory to set up the European Physiology Module. He examined the hatches of the U. S. segment. Gerst updated the software on two laptop computers, installed alignment guides in the Combustion Integrated Rack, then sanitized and inspected hatches. Later, he recorded the magnetic field strength inside the European Drawer Rack (EDR) using a teslameter before installing a new experiment. He then installed the Magvector in the EDR to investigate how Earth's moving magnetic field interacts with an electrical conductor. Suraev searched for microbes growing in the Zvezda service module and later audited medical kits and maintained the Elektron oxygen generator in Zvezda. Skvortsov took blood and saliva samples and Artemyev photographed the inside of the Russian segment.

Wiseman and Swanson spoke with students from Elliot Ranch Elementary School in Elk Grove, California on August 27. Gerst worked on the VIABLE microbiology experiment in the Zarya cargo module checking experimental materials in a locker panel. He then carried out maintenance on the network infrastructure in the Columbus module. He photographed windows of the copula with the shutters closed. This was to check if hardware would interfere with upcoming IMAX filming. Skvortsov and Artemyev spoke with ground controllers about their upcoming landing in Kazakhstan and the rescue operations.

The ATV Georges Lemaître docked to the Zvezda and fired its thrusters to raise the station's orbit in preparation for the arrival of Soyuz TMA-14M due in September.

Swanson worked on the Robonaut upgrade on 28 August 28. He installed the new legs and the associated gears and cables. He also prepared the Advanced Colloids Experiment (ACE) hardware for further experiments. Wiseman and Gerst gave each other eye examinations, capturing detailed images of their retinas. Wiseman stowed EVA equipment to make room for future installations. He photographed the snare cables on the Canadarm2's Latching End Effector (LEE) from the cupola. In the Columbus laboratory, he worked on the Biolab, exchanging filters sponges. Skvortsov took blood and saliva samples. The Immuno study investigates saliva, blood and urine samples to observe changes in a crew member's stress and immune responses. Suraev and Artemyev installed overlay sheets on interior panels in the Zarya cargo module. Suraev also checked air flow sensors and took inventory in the Russian segment.

Gerst uninstalled rendezvous equipment that was not needed after the arrival of the Georges Lemaître on August 29. He set up the Energy experiment in the ESA Columbus laboratory. This French study looks at nutrition as a way to maintain crew energy balance, health and performance on long duration space missions. Wiseman took air samples in the U. S. segment with the Microbial Air Sampler and petri dishes. He stowed the petri dish samples in bags for later analysis.

Expedition 40, Week Ending September 7, 2014

Swanson started the week setting up acoustic dosimeters to be worn by himself and Suraev for 24 hours at a time. He then tested the skin on his forearm for the Skin B experiment. Afterward, he carried out leak tests of the Sokol launch and entry suits with Skvortsov and Artemyev in their Soyuz TMA-12M. Wiseman unloaded pre-configured hard drives from the ATV and installed them in laptops. The ground controllers transferred control from one set of computers to another. Suraev and Gerst set up Matryoshka bubble dosimeters in the Harmony node to measure radiation as part of the RaDI-N study. Afterward, Gerst spoke to RTL-TV in Cologne, Germany, giving them the story of his mission so far. After lunch, Swanson set up the Portable Pulmonary Function System hardware for the Sprint VO2max sessions, to be conducted by himself and Wiseman. On this day the ground controllers configured an artificial 50 seconds delay in the communication link with Swanson. This is part of the Communications Delay Assessment experiment that simulates the delay in communications between Mission Control and a deep space mission. Wiseman continued to install new computer hard drives and configured another run of the Binary Colloidal Alloy Test-Kinetics Platform. Wiseman and Swanson also spoke to students at the INFINITY Science Center, the visitor center associated with NASA's Stennis Space Center, Mississippi.

On September 3, Gerst and Artemyev took the acoustic dosimeters previously worn by Swanson and Suraev. They wore them for a further 24 hours to learn about the sound levels to which the crew is exposed. Swanson and Skvortsov then packed up their remaining crew provisions to be loaded into the Soyuz TMA-12M. Suraev and Artemyev recharged the satellite phone to be packed in the Soyuz for communications once landed. Artemyev then worked on the SPLANH study which takes a look at the effects of spaceflight on the digestive system. Skvortsov installed the HD camera in the Soyuz to capture video of the descent. Gerst wore a breathing mask in the morning and relaxed as much as possible while recording his oxygen uptake for the ENERGY experiment. Wiseman gathered water samples for the Microbiome study. Swanson then joined him in collecting surface samples to find microbial contamination. Swanson analysed the samples.

Wiseman talked with students in Evansville, Indiana, over the station's ham radio. After Gerst downloaded the data from his ENERGY relaxation session, he spent the remainder of the day working out on the station's exercise bike and ARED.

Wiseman placed acoustic dosimeters in the Destiny laboratory and the Zarya module on September 4 to determine the noise levels the crew is exposed to throughout the day. He then tested samples from the station's Water Processor Assembly to check for signs of contamination. Gerst stowed the hardware he used the previous day to measure his oxygen uptake for the Energy experiment. Skvortsov and Artemyev packed and stowed items inside the Soyuz-TMA-12M spacecraft as the two cosmonauts prepared to depart the station along with Swanson. Artemyev also activated the Membrana experiment hardware, which is testing the capability of using the weightless environment of the station to produce porous materials with uniform characteristics. Suraev, focused on maintenance tasks as he checked cable connections, cleaned filter screens and checked the life-support system hardware in the Zvezda service module. Gerst, who spent part of the afternoon setting up the Kubik incubator/cooler, took a break from his work to talk to his hometown of Künzelsau, Germany, during a televised ESA event. He then collected detailed imagery of Wiseman's eyes using a fundoscope. Suraev spent the afternoon photo-documenting the interior of Zvezda while Skvortsov and Artemyev continued preparing the Soyuz for the journey back to Earth.

In preparation for their return to Earth, Skvortsov and Artemyev tested the motion control system of the Soyuz TMA-12M spacecraft on September 5. They spent the remainder of the morning conducting Lower Body Negative Pressure training. Swanson, Wiseman, Gerst and Suraev all began the day with medical specimen collections, providing saliva and blood samples for various experiments that track the effects of long-duration spaceflight on the human body. After Swanson stored his blood sample in the MELFI, he noticed that the doors on the NanoRacks cubesat deployer mechanism were open. That mechanism, which was attached to the end of the Japanese robotic arm on the exterior of the Kibō module, was designed to eject cubesats into orbit. Flight controllers determined that two cubesats had been inadvertently deployed yet again. No crew members or ground controllers saw the deployment, and no views of the deployment were found on the video footage recorded by the station's cameras. The deployment appears to have occurred sometime overnight or possibly hours after Japanese flight controllers "jiggled" the robotic arm in an effort to get the doors to open. The deployer mechanism had been out on the arm since the previous month. Wiseman and Gerst carried out computer-based training for the upcoming capture of the SpaceX Dragon cargo vehicle. Meanwhile, Suraev collected surface samples in the Russian segment of the station for research into the biodegradation of the surfaces in materials used in station construction. Russian researchers were looking into the kinds of microorganisms that may be colonizing those surfaces to determine the best methods for preventing corrosion and damage.

Expedition 40/4, Week Ending September 14, 2014

Swanson, Wiseman and Gerst assembled in the Destiny laboratory on September 8 for an interview with KOA Radio in Denver, Colorado. Wiseman then floated over to the Harmony node and conducted a show-and-tell with the *Wall Street Journal*. Swanson handed over his spacewalk duties to Gerst. Swanson and Wiseman were scheduled to perform a pair of spacewalks in August. These were postponed until after the SpaceX mission due to launch on September 19, which contains new spacesuit equipment. This date would be after Swanson's departure so Gerst and Wiseman were now assigned the two spacewalks planned for October. Skvortsov and Artemyev continued packing equipment inside the Soyuz for the ride home. They also worked on various maintenance tasks, including charging batteries, filling an oxygen generator tank and sampling for microbes in the station's Russian segment. Suraev worked on the Matryoshka radiation study. He then moved on to a calcium study that observes mineral loss in crew members' bones. Finally, he switched over to the Interaction study, which studies how crew members of various cultural and international backgrounds work together during different phases of a space mission.

Gerst installed a microscope for the JAXA Cell Mechanosensing-2 experiment in the Kobairo rack on September 9. This experiment tries to identify gravity "sensors" in human cells that may change the expression of key proteins and genes and allowing muscles to atrophy in microgravity. Suraev worked on a radiation exposure experiment. The crew continued to swap the acoustic dosimeters every 24 hours. Skvortsov and Artemyev saw their science and maintenance schedule get lighter as they neared the end of their stay on the station. Artemyev sampled air in the Zvezda service module for ammonia and monitored its sanitary and epidemiological status. Skvortsov and Suraev gathered the Matryoshka detectors and stowed them in the Soyuz for return to Earth. Swanson took and stored a urine sample, then packed the medical kit and cleaned port-side crew quarters before maintaining the Combustion Integrated Rack.

Swanson handed over command of the International Space Station to Suraev on September 9 at 5:15 p.m. EDT in a traditional Change of Command Ceremony. Alexander Skvortsov, Expedition 40 commander Steve Swanson and Oleg Artemyev undocked the Soyuz TMA-12M spacecraft from the Poisk mini-research module at 7:01 p.m. EDT on September 10. This marked the end of Expedition 40 and the start of Expedition 41. The Soyuz touched down with a parachute-assisted landing less than an hour later southeast of the town of Dzhezkazgan, Kazakhstan at 10:23 p.m. EDT.

With Max Suraev in command of Expedition 41, the three-person crew had a day off on September 11 after a full day of undocking activities. Gerst photographed the Soyuz as it was separating from the station. Wiseman captured a photograph of the Soyuz and its plasma trail the moment it re-entered the atmosphere. The crewmates were expecting three new colleagues to arrive on September 25, with Alexander Samokutyaev, Barry Wilmore and Elena Serova expected to join Expedition 41.

Wiseman and Gerst took turns with the ENERGY experiment when they returned to duty on Friday, 12 September. Wiseman also worked throughout the day on routine upkeep inside the International Space Station. He first replaced a cable on the advanced resistive exercise device (ARED) then performed light plumbing working on the Water Recycling System. After that he cleaned filters inside the Microgravity Science Glovebox before lunch time. Gerst activated a SpaceX UHF communications unit ahead of Dragon's arrival. The device allows the crew to send basic commands to the private space freighter as it approaches the space station. He also assisted Wiseman at the beginning of the day with the ARED cable replacement work. He then switched to light computer work before filling out a weekly questionnaire documenting headaches in space. In the after-noon, Gerst photographed CubeSat hardware before moving on to Crew Medical Officer computer-based training. Suraev spent his day primarily inside the sta-tion's Russian segment working on his complement of science and maintenance. First, he configured communications equipment in the Rassvet mini-research module. Then he copied science data to laptop computer. He then worked in the Zvezda service module for preventive maintenance on the ventilation system.

Expedition 41, Week Ending September 21, 2014

Suraev, Wiseman and Gerst checked their Soyuz Kazbek seat liners in the Soyuz TMA-13M spacecraft docked at the Rassvet Mini-Research Module-1. Suraev maintained and cleaned the ventilation system in the Zvezda service module. Wiseman meanwhile maintained the Waste and Hygiene Compartment in the Tranquility node. Then he prepared the vestibule of the Harmony node for the arrival of the SpaceX Dragon cargo ship. At the same time, Gerst prepared the Robotics Work Station in the cupola and the backup workstation in the Destiny laboratory. This was to prepare for the robotic grappling of the cargo ship. Gerst also tested the recycled water using the Total Organic Carbon Analyzer. Later, he recorded the noise levels in the station using a sound level meter. Suraev worked with the Vizir experiment, which is designed improve the targeting of Earth pho-tography by cosmonauts through the use of ultrasonic angle measurements.

The thrusters of the Georges Lemaitre ATV were used to raise the station's orbit on Saturday night. The 3 minutes, 44 seconds firing raised the perigee of the sta-tion's orbit by 1.2 statute miles, leaving the station in a 262.8 × 253.9 mile orbit and ready for the arrival of Soyuz TMA-14M.

The new station commander, Suraev began September 16 checking the hard-ware for the Otklik experiment. This tracks the impacts of particles on the station's exterior using piezoelectric sensors. He spent the remainder of the day focused on maintenance activities in the Russian segment of the station as he made sure the ventilation system in the Zvezda service module remained in good working order and checked up on the SOZh life-support system. Wiseman analysed a urine sample for the Pro K experiment and conducted proficiency training to maintain his Crew Medical Officer rating. Later, he installed alignment guides inside the Combustion Integrated Rack to support more ground-commanded experiments into the behaviour of ignited fuels in microgravity. Gerst spent much of his day in the Japanese Kibō laboratory as he prepared that module's airlock for the return of the NanoRacks CubeSat deployer mechanism back inside the station. The deployer mechanism was mounted at the end of the Japanese robotic arm on the exterior of the Kibō module. In the previous month it successfully deployed some CubeSats but failed to deploy others. Gerst set up the airlock's slide table to receive the deployer mechanism and placed a protective cover over the airlock's window.

Gerst and Wiseman prepared for the upcoming SpaceX CRS-4 cargo ship, the fourth commercial resupply mission. This was scheduled to launch the following Saturday. They held a conference call with Mission Control and conducted on-board training with the Canadarm2. Wiseman then packed up the Fundamental and Applied Studies of Emulsion Stability experiment (FASES), for return to Earth aboard Dragon. He also initiated the Commercial Generic Bioprocessing Apparatus to control the environment for experiments on cells, microbes and plants. Gerst pressurized the Kibō's scientific airlock and accessed the NanoRacks CubeSat deployer mechanism inside the airlock. He secured the doors to the two remaining cubesats to prevent any unplanned launches. Suraev loaded rubbish into the Progress 56 cargo craft. All three crewmembers ended the shift talking to their replacements on the ground. They transmitted "lessons learned" during their stay on the ISS.

Wiseman and Gerst prepared the Quest airlock for two U. S. spacewalks scheduled for October. They set up tools and equipment needed for the EVA. Wiseman participated in the bone density Pro K study. On Sunday, Wiseman and Gerst reviewed the procedures for grappling Dragon once again had a last Canadarm2 robotics training session with the Robotics Onboard Trainer.

The SpaceX Dragon spacecraft lifted off on the Falcon 9 rocket from Cape Canaveral Air Force Station in Florida at 1:52 a.m. EDT September 21, carrying about 5000 pounds of cargo. The mission is designated CRS-4 under the Commercial Resupply Service programme and SpX-4 under the ISS flight numbering system. This was a day later than originally scheduled due to adverse weather at the launch site.

Expedition 41, Week Ending September 28, 2014

Wiseman and Gerst continued to train and prepare for the robotic grappling and berthing of the SpaceX capsule on September 22. Gerst also installed camera equipment in the Multipurpose Small Payload Rack and the Aquatic Habitat in the Kibō laboratory. This was in preparation for the Zebrafish Muscle experiment.

Suraev carried out routine maintenance tasks in the Russian segment. He started in the Poisk docking compartment and ended in the Zarya cargo module after cleaning ventilation grilles and replacing dust filters.

Wiseman and Gerst commanded the Canadarm2 from the robotics workstation in the cupola to grapple the Dragon at 6:52 a.m. EDT September 23. They opened the Dragon hatch a day ahead of schedule and entered the vehicle. The Dragon arrived with 2216 kg of crew supplies, hardware, experiments, computer equipment and spacewalk equipment. This was the fourth SpaceX mission under the Commercial Resupply Services contract.

On September 24, Gerst transferred a science freezer from Dragon for installation in an EXPRESS rack inside the Destiny laboratory. The freezer, known as General Laboratory Active Cryogenic ISS Experiment Refrigerator (GLACIER), stores scientific samples at temperatures ranging from +4°C (39° F) to −160° C (−301° F). He relocated lockers in an EXPRESS rack that would house the Rodent Research hardware. Gerst also prepared samples to start incubation for the Cell Mechanosensing experiment and he transferred more cellular research equipment for the Micro-8 study from Dragon in the Destiny laboratory. He inserted the equipment into a Commercial Generic Bioprocessing Apparatus inside an EXPRESS rack. Wiseman performed maintenance on the COLBERT treadmill inside the Tranquility node. He swapped computers from the Cupola and Tranquility to troubleshoot a connection to the COLBERT. Later, he photographed the configuration of Dragon as Gerst entered the vehicle to begin cargo transfers. They also reviewed emergency procedures necessary while Dragon is attached to the station. Wiseman then unpacked Double Coldbags from Dragon that are part of science freezer hardware. Finally, he and Gerst got back together for a conference with payload specialists to discuss operations with the Rodent Research-1 hardware. Suraev worked on three different science experiments. He started with the VIZIR study which seeks to improve Earth observation photography techniques. He then moved on to Virtual for work to determine how a crew member's vestibular system adjusts to weightlessness. Finally, he checked hardware for the Otklik experiment that is studying ways to detect and record micrometeoroid impacts on the International Space Station. At the end of the day, he conducted a ham radio session with students from Quito, Ecuador.

NASA astronaut Butch Wilmore and Russian cosmonauts Alexander Samokutyaev and Elena Serova launched from the Baikonur Cosmodrome in Kazakhstan at 4:25 p.m. EDT September 24. They launched aboard the Soyuz TMA-14M spacecraft, docked at the Poisk zenith docking station and the hatches were opened at 1:06 a.m. EDT. Expedition 41 Commander Max Suraev of the Russian Federal Space Agency and Flight Engineers Reid Wiseman of NASA and Alexander Gerst of the European Space Agency welcomed the new crew members aboard their orbital home.

During the ascent and orbits, the port solar array of the Soyuz did not deploy. This posed no threat to the crew or the mission. Shortly after docking, it deployed. NASA and Roscosmos officials confirmed that the array posed no long-term issue to either standard operation at the station for Expedition 41-42 or for the landing of Wilmore, Samokutyaev and Serova at the conclusion of their mission in March.

For Samokutyaev and Wilmore, this was their second spaceflight and visit to the station. Samokutyaev was a member of Expeditions 27 and 28 in 2011. Wilmore visited the station on the space shuttle *Atlantis* during the STS-129 mission. It was Serova's first spaceflight and she became Russia's first, and at the time of writing only, female long-term crew member.

There were now five spacecraft docked to the station, its maximum visiting vehicle capacity: two Soyuz, one Progress, the ATV-5 and the SpaceX Dragon.

Expedition 41, Week Ending October 5, 2014

With a full crew of six onboard, the crew returned to work. Wiseman, Gerst and Wilmore, prepared for upcoming spacewalk activities on September 29. Wiseman and Gerst reviewed the procedures for the EVA and held a conference call with spacewalk specialists in Mission Control. Wilmore scrubbed Liquid Cooling and Ventilation Garment in his spacesuit. Gerst checked water quality of the tank in the Zebrafish Muscle study. Serova transferred test samples from the Kaskad cell cultivation experiment, while Samokutyaev investigated aseptic hardware intended for use in biotechnological experiments. Suraev packed away rubbish in the Progress 56 to burn up on re-entry. The entire crew then met to review their roles and responsibilities during an emergency aboard the station. Wilmore, Samokutyaev and Serova had a planned hour for familiarization with the station.

Mission Control robotic controllers maneuvered the Canadarm2 robotic arm and its Special Purpose Dexterous Manipulator to the trunk of the recently arrived SpaceX Dragon cargo ship attached to the Earth-facing port of the Harmony node. They grappled and removed an adapter mechanism from Dragon's trunk and installed it to a bracket on the front of the Columbus module. This was in preparation for the removal and installation of the ISS-Rapid Scatterometer, or RapidScat, the following day. RapidScat monitors ocean wind speed and direction using radar.

Gerst transferred test samples for the Micro-8 experiment on September 30. This experiment investigates the Candida albicans yeast in order to help scientists better understand and control the infectious nature of this opportunistic pathogen. Wilmore set up equipment inside the Microgravity Science Glovebox associated with the Rodent Research hardware system, which provides a platform aboard the station for long-duration rodent experiments in space. Suraev checked out a spare cable for a control panel, updated procedure documents and performed routine maintenance on the life-support system in the Zvezda service module. Samokutyaev continued work with the Aseptik experiment, which is testing methods and equipment for maintaining the sterility of hardware used for biotechnology studies aboard the station. Serova meanwhile mixed some new test sample for the Kaskad cell cultivation experiment.

Wiseman, Wilmore and Gerst spent October 1 in the Quest airlock resizing their U. S. spacesuits for a pair of spacewalks. They collected measurements of their bodies to compare with baseline measurements taken before launch, and then they put on their spacesuits to make sure everything fit properly. These on-orbit fit checks are necessary because astronauts may grow up to 3% taller while living aboard the space station. Wiseman and Gerst were to perform the first EVA and Wilmore was to replace Gerst for the second. Wiseman and Gerst also reviewed operating procedures for the Simplified Aid For EVA Rescue (SAFER). This is a backpack with small nitrogen-jet thrusters to propel the astronaut back to safety in the event of being untethered. Wilmore also prepared seed samples and a culture dish for the Plant Gravity Sensing experiment, which examines the cellular and molecular mechanisms that enable plants to sense gravity. Wiseman meanwhile checked in on the Rodent Research experiment, which looks at how living in space affects rodents and how that knowledge might be applied to humans. Suraev performed an equipment check for the Otklik experiment, which tracks the impacts of particles on the station's exterior. He then gathered data from the Matryoshka radiation-detection study before moving on to stow trash and unneeded items into the Progress 56. Samokutyaev wrapped up his work with the Aseptik experiment, which is testing methods and equipment for maintaining the sterility of hardware used for biotechnology studies aboard the station. Serova worked with the Kaskad cell cultivation experiment throughout the day, manually mixing test samples within its bioreactor. She also removed the lights and cameras from inside the Soyuz TMA-14M for return to Earth aboard the Soyuz TMA-13M spacecraft.

Gerst took samples from the Commercial Generic Bioprocessing Apparatus and injected test drugs into them on October 3. The Drug Metabolism experiment investigates whether a drug used to treat Type 2 diabetes (metaformin) can be used to treat cancer. He also transferred a seedling culture dish from the Plant Gravity Sensing experiment into the Cell Biology Experiment Facility for incubation. This experiment investigates how plants use gravity to determine direction of light and nutrients. Wilmore maintained the Aquatic Habitat in the Kibō laboratory. Wiseman put samples from the Biological Research in Canisters-19 (BRIC-19) experiment into the MELFI freezer. He also transferred test samples for the Micro-8 experiment. Samokutyaev used the Cardiovector health experiment to see how his heart was adapting to microgravity and long-duration spaceflight. Serova mixed test samples within the bioreactor of the Kaskad cell cultivation experiment. Later she photographed and deployed new samples for the calcium bone loss experiment.

Expedition 41, Week Ending October 12, 2014

Wiseman and Gerst reviewed their EVA timeline and prepared the Quest airlock on October 6. They also checked out their SAFER units. Wilmore assisted the pair and also trained for his role as the robotic arm operator for the EVA. Gerst also checked the water quality in the Aquatic Habitat. Samokutyaev and Serova continued to familiarize themselves with station systems and safety procedures. Samokutyaev joined Suraev to review the plans of their next Russian spacewalk.

Wiseman and Gerst completed the first of three Expedition 41 spacewalks at 2:43 p.m. EDT on October 7. The EVA lasted 6 hours and 13 minutes. They relocated a failed cooling pump to External Stowage Platform-2 (ESP-2) outside the Quest airlock. They stowed adjustable grapple bars on ESP-2. Gerst replaced a light on an External Television Camera Group (ETVCG) outside Destiny. They also installed a Mobile Transporter Relay Assembly (MTRA) on to the S0 truss above the Destiny laboratory. The MTRA adds the capability to provide "keep-alive" power to the Mobile Servicing System when the Mobile Transporter is moving between worksites.

Fig. 4.10. Alexander Gerst during the first EVA, October 2014 (NASA)

Gerst and Wiseman had a half day rest on October 9 after their spacewalk exertions. Once back on duty, Wiseman set up power and data cables for the Cygnus hardware command panel. Gerst, in the Columbus laboratory, contacted students from two schools in the United Kingdom by ham radio. Wilmore joined Wiseman and Gerst for a spacewalk debrief with flight controllers on the ground. Serova began her work day collecting dosimeter readings in the station's U. S. segment as part of post-spacewalk procedures. She then floated back to the Rassvet module to clean fan screens. Then she moved on to crew orientation activities before some plumbing work transferring fluids to a tank in the Pirs docking compartment.

Wilmore and Wiseman reviewed the new Rodent Research-1 experiment on October 9, and then carried out the Cardio Ox study that uses an ultrasound to scan a crew member's carotid and brachial arteries. They joined Suraev just before lunch time for a television conference with Russian schoolchildren and the leader of Roscosmos. All six crew members gathered together for an emergency drill. They simulated an emergency to practice communication with ground controllers, locate safety equipment and to familiarize themselves with escape paths.

Samokutyaev and Suraev prepared for their upcoming spacewalk, checked their Orlan spacesuits and spacewalk tools and reviewed their tasks. Serova collected air samples to check the air quality in the station's Russian segment and she worked on the long-running Matryoshka radiation absorption experiment.

Expedition 41, Week Ending October 19, 2014

Wiseman and Wilmore prepared on October 14 for their EVA. They configured their spacesuits in the Quest airlock. Suraev and Samokutyaev performed leak-checks om their Russian Orlan spacesuits for their EVA. They visually reviewed the worksites and travel routes. They also tested the Telerobotically Operated Rendezvous Unit (TORU), in the Zvezda service module in preparation for the undocking of the Progress 56 cargo ship and the docking of the Progress 57. Serova checked out the Otklik experiment and she mixed test samples in the Kaskad cell-cultivation experiment's bioreactor. Gerst maintained the Aquatic Habitat and activated test samples for the Micro-8 study inside the Commercial Generic Bioprocessing Apparatus-6.

Wiseman and Wilmore started their spacewalk by switching their suits to battery power at 8:16 a.m. EDT, October 15. After exiting the Quest airlock, they translated out to the starboard side of the station's integrated truss structure where replaced a failed power regulator, the sequential shunt unit. This had stopped working in May. Since then, the used only seven of its eight power channels. They had to wait until the station was in the Earth's shadow before they replaced the unit. During this time the solar array was not "live". The finished the EVA by moving some equipment on the port side of the station's truss in preparation for the relocation of the *Leonardo* Permanent Multipurpose Module (PMM). Wiseman and Wilmore replaced a faulty TV camera on the P1 truss segment. They also detached an articulating portable foot restraint and tool stanchion from P1 and moved it inward to the centrepiece of the station's truss structure, the S0 truss. Finally, they moved the Wireless Video System External Transceiver Assembly (WETA) from the P1 truss to the top of the Harmony node. The 6 hours, 34 minutes spacewalk ended at 2:50 p.m. EDT. It was the 183rd EVA in support of station assembly and maintenance.

Wiseman and Wilmore took some time off on October 16, then they participated in some post-spacewalk health exams, performed maintenance on the spacesuits and stowed some of the equipment. Wilmore also spent some time activating
and mixing samples for a NanoRacks experiment. Gerst carried out maintenance
of the Oxygen Generator System, which provides breathing air for the station's
crew. Suraev and Samokutyaev checked out the communication system of their
Orlan spacesuits. Later, they used the Dynamic Onboard Ubiquitous Graphics
application to review the worksites for their spacewalk and the paths they'll need
to take to on the exterior of the station reach those locations. Suraev also closed
the hatch to the Progress 56 and conducted a leak check of the seal around the
hatch. Meanwhile, Serova performed routine daily maintenance on the life-support
system in the Zvezda service module.

Suraev and Samokutyaev, with assistance from Serova, transferred a pair of
pressurized Orlan spacesuits to the airlock and reviewed emergency procedures.
Samokutyaev and Suraev also installed the spacesuit helmet cameras and lights on
loan from their NASA crewmates and finished up the day with a review of airlock
procedures. Serova also manually mixed test samples in the Cascade cell-
cultivation experiment's bioreactor.

Expedition 41, Week Ending October 26, 2014

ISS commander Suraev and Samokutyaev spent October 20 readying the Russian
Orlan spacesuits for their spacewalk. They put on the spacesuits and checked suit
equipment without leaving the station and carried out leak checks.

The departure of SpaceX CRS-4 was postponed to October 25, due to high seas
caused by storms in the splashdown zone exceeded recovery rules.

Serova activated equipment for the Cascade biological experiment which investigates cell cultivation. She also cleaned fans and filters and updated the inventory
management system.

While Suraev and Samokutyaev reviewed their spacewalk plan one final time
and completed the setup of the Pirs airlock, Gerst helped ready the Russian segment for the spacewalk when he closed the hatch to the ATV-5. Gerst then replaced
the water in the Aquatic Habitat in Kibō. Wilmore and Wiseman worked in the
Quest airlock cleaning up after a pair of spacewalks conducted over two weeks.
They took turns scrubbing the cooling loops on the U. S. spacesuits and collecting
water samples from the loops. Serova worked throughout Tuesday helping Suraev
and Samokutyaev get ready for Russia's 40th spacewalk at the ISS. She also continued more dust filter replacement and fan screen cleaning activities.

Max Suraev and Alexander Samokutyaev conducted a Russian segment EVA
on October 22. The Radiometriya experiment had been mounted on the exterior
of the Zvezda service module in 2011. As it was no longer needed, they uninstalled it and manually jettisoned it, to burn up in re-entry, when its orbit decayed.
It had recorded data to predict earthquakes. They then photographed the ESA

Expose-R experiment which exposes organic and biological samples to the harsh environment of space and observes how they are affected by cosmic radiation, vacuum and night and day cycles. They moved to the Poisk mini-research module and removed a pair of rendezvous antennas no longer needed that were blocking translation paths for future spacewalks. These were also manually jettisoned. Finally, the pair photographed the exterior surface of the Russian modules. In the past the configuration of the Russian segment was not documented clearly and caused problems during spacewalks. The Russians were making a detailed model of the exterior.

This was the second spacewalk for both cosmonauts and closed the Pirs docking compartment hatch at 1:06 p.m. EDT ending the third spacewalk for Expedition 41. The EVA lasted 3 hours and 38 minutes. It was the 184th in support of station assembly and maintenance.

On October 24 the hatch of the SpaceX Dragon was closed for the final time. It was unberthed on October 25 and released by the Canadarm2 at 9:57 a.m. EDT. It splashed down in the Pacific Ocean at 3:39 p.m. EDT a few hundred miles west of Baja California, Mexico, with 3276 pounds of cargo and science samples. Dragon capsules are the only uncrewed spacecraft that return to Earth and do not burn up on re-entry.

Expedition 41, Week Ending November 2, 2014

At 1:38 a.m. EDT October 27, the Progress 56 cargo spacecraft separated from the station and fired its engines to move away. It was planned to carry out three weeks of engineering tests before deorbiting the craft over the Pacific. This cleared the Pirs docking compartment for the arrival of the new Progress 57 resupply spacecraft.

The launch of the Orbital Sciences' Antares rocket carrying its Cygnus cargo spacecraft was scrubbed on Monday as there was a boat down range in the trajectory.

On October 28 the Cygnus CRS Orb-3 and its Antares rocket exploded 15 seconds after launch.[5]

On October 29, at 3:09 a.m. EDT, the Progress 57 cargo craft launched from the Baikonur Cosmodrome in Kazakhstan. Traveling about 261 miles over the Atlantic Ocean, the unpiloted Progress 57 cargo ship docked at 9:08 a.m. EDT to the Pirs Docking Compartment. It contained almost 3 tons of food, fuel and supplies, including 880 kg of propellant; 22 kg of oxygen; 26 kg of air; 420 kg of water; and 1280 kg of spare parts, supplies and experiment hardware.

---

[5] It was later determined that a turbo pump failed on an Aerojet Rocketdyne AJ-26 engine, which was a refurbished Russian NK-33 engine. This was the second cargo craft to fail to reach the station. Progress 44P failed to reach orbit in 2011. In 2015, a further Progress and SpaceX's Dragon CRS-7 would also fail to reach the station, highlighting the dangers of spaceflight.

Suraev and Samokutyaev opened the hatch to the Progress 57 on October 30 and began unloading its cargo. Suraev also joined Wiseman for descent training in advance of their return to Earth. Wilmore and Gerst scrubbed cooling loops in the U. S. spacesuits throughout the day. Gerst also changed the water in the Kibō laboratory's Aquatic Habitat.

Wiseman, Gerst and Suraev checked their Sokol suits on October 31. Wilmore worked on the Zebrafish Muscle experiment in the Kibō module. Serova participated in Crew Medical Officer training, and then moved on to a variety of science and maintenance tasks. Samokutyaev worked on cargo transfers from the Progress 57 and performed preventative maintenance in the Pirs docking compartment.

The crew noted that this was 14th anniversary of the launch of Expedition 1, the first crew to live and work aboard the International Space Station.

Expedition 41, Week Ending November 9, 2014

November 3, 2014, was the 14th anniversary of the docking of the Soyuz TM-31 spacecraft at the ISS. The space station at the time consisted of just three modules: the Unity node, the Zarya cargo module and the Zvezda service module. Commander William Shepherd and Flight Engineers Sergei Krikalev and Yuri Gidzenko became the first crew to live on the station.

Gerst drew his blood samples for stowage in a science freezer and he also worked on the Zebrafish Muscle experiment in the Kibō laboratory. Wiseman worked on plumbing tasks, then set up cameras for the Sally Ride EarthKAM experiment. Wilmore checked for leaks and worked on a fan in a U. S. spacesuit. Serova sampled the station's air and surfaces for microbes. Samokutyaev transferred cargo from the Progress 57 and assisted Wiseman with the EarthKAM study.

While preparing to return to Earth, Gerst wrote in his ESA blog:

I owe my life at this point to five letters: ECLSS. Of course, I owe my life to many things, but floating up here in the vacuum of space it is our life-support system that takes a starring role in providing us with the basics of living: oxygen, water and a comfortable ambient temperature.

The space industry loves Three-Letter Acronyms (TLA) so the fact that the ECLSS has five letters is a testament to its importance for Max, Reid, Elena, Butch, Alex and me. It stands for Environment Control & Life Support Systems.

Note the plural "Systems." Not one unit, but multiple machines spread out over the International Space Station to create oxygen for us to breathe and remove the carbon dioxide that we breathe out. The ECLSS recycles water from condensation, sweat and even our urine and turns it into drinking water.

Centrifuges spin to separate the water from gasses while filters remove contaminants. The system does the same as our ecosystem on Earth: recycling waste water to turn it into fresh water. The result is chemically pure, perfectly drinkable, and tastes great.

Our ECLSS does even more, though; it keeps us warm or cool as needed, circulates our air, monitors its quality and gives us an early warning in case of noxious gas leaks or fire.

Unfortunately, the system is not completely self-sustainable yet. Electrolysis is used to separate oxygen atoms from water, while the extra hydrogen is combined with carbon dioxide to make water again. The resulting methane is vented out into space. Regular supply ships such as the Progress, Dragon or ATV make up for lost molecules when they bring fresh supplies of oxygen, water and food.

I have used ESA's space freighter ATV-5 that is now docked to our home to add some oxygen to our living space. I pressurized the Space Station by around 0.015 bar (12 mmHg) of pure oxygen.

The systems are not perfect yet and require maintenance often, so we spend quite some time working with them, such as cleaning filters, exchanging pumps and balancing out fluids in the different tanks. But every day we learn more about how to run these systems in space, and get a step closer to knowing how to build the life support system that will sustain a mission to Mars and beyond. Of course, this knowledge also helps us to build much more efficient waste treatment plants for use on Earth.

It was not until 2009 that the International Space Stations life support could handle more than three astronauts for longer periods. Working on the ECLSS I often think about how our beautiful planet silently supplies 7 billion people with what they need to stay alive.

As a single human being on Earth it is hard to encompass the intricate balance of organisms and processes that turn one beings waste into another's source of life. Seeing this ecosystem from above – our pale blue dot – it is still unfathomable.

But one thing does become clear: our ecosystem is not a silent all-encompassing never-ending resource, but a fragile, thin layer of life-support on what would otherwise be a world as barren as the Moon.

We need to take care of our life-support system on Earth, just as we need to maintain the ECLSS up here. Recycling, care and attention go a long way.

Soyuz Commander Max Suraev and Flight Engineers Reid Wiseman and Alexander Gerst spent the morning of November 6 reviewing their Soyuz undocking and descent activities ahead of their landing in Kazakhstan.

See Appendix "Alexander Gerst Blog on Science" for Alexander Gerst's blog entry describing the science of the Blue Dot mission. It was published on the ESA website on November 6, 7 and 8. 2014.

At 4:27 p.m. EST, November 9, the hatches were closed between the International Space Station and Soyuz TMA-13M, containing Expedition 41 crew members Reid Wiseman, Alexander Gerst and Soyuz Commander Maxim Suraev. The Soyuz undocked at 7:31 p.m. They landed their spacecraft in Kazakhstan at 10:58 p.m. EST, having spent 165 days aboard the space station. This was the first mission for both Wiseman and Gerst. Suraev had by now spent 334 days in space during two missions.

Postscript

Fig. 4.11. Alexander Gerst, German Chancellor Angela Merkel and Thomas Reiter at Stralsund, Germany, 2015 (ESA)

Gerst is currently carrying out post-flight activities at ESA's Astronaut Centre in Cologne, Germany. After his Blue Dot and Horizon missions, he has spent 362 days 1 hours and 51 minutes in space, making him the current record holder for time in space of any ESA astronaut, active or retired.

5

Futura

Mission

ESA Mission Name:	Futura
Astronaut:	Samantha Cristoforetti
Mission Duration:	199 days, 16 hours, 42 minutes
Mission Sponsors:	ASI/ESA
ISS Milestones:	ISS 41S, 78th crewed mission to the ISS

Launch

Launch Date/Time:	November 23, 2014, 21:01 UTC
Launch Site:	Pad 31, Baikonur Cosmodrome, Kazakhstan
Launch Vehicle:	Soyuz TMA
Launch Mission:	Soyuz TMA-15M
Launch Vehicle Crew:	Anton Nikolayevich Shkaplerov (RKA), CDR
	Samantha Cristoforetti (ESA), Flight Engineer
	Terry Wayne Virts (NASA), Flight Engineer

Docking

Soyuz TMA-15M	
Docking Date/Time:	November 24, 2014, 02:48 UTC
Undocking Date/Time:	11 June 2015, 10:20 UTC
Docking Port:	Rassvet nadir

Landing

Landing Date/Time:	June 11, 2015, 10:20 UTC
Landing Site:	near Dzhezkazgan, Kazakhstan
Landing Vehicle:	Soyuz TMA
Landing Mission:	Soyuz TMA-15M
Landing Vehicle Crew:	Anton Nikolayevich Shkaplerov (RKA), CDR
	Samantha Cristoforetti (ESA), Flight Engineer
	Terry Wayne Virts (NASA), Flight Engineer

© Springer Nature Switzerland AG 2020
J. O'Sullivan, *European Missions to the International Space Station*,
Springer Praxis Books, https://doi.org/10.1007/978-3-030-30326-6_5

ISS Expeditions

ISS Expedition:	Expedition 42
ISS Crew:	Barry Eugene Wilmore (NASA), ISS-CDR
	Aleksandr Mikhailovich Samokutyayev (RKA), ISS-Flight Engineer 1
	Yelena Olegovna Serova (RKA), ISS-Flight Engineer 2
	Anton Nikolayevich Shkaplerov (RKA), ISS-Flight Engineer 3
	Samantha Cristoforetti (ESA), ISS-Flight Engineer 4
	Terry Wayne Virts (NASA), ISS-Flight Engineer 5
ISS Expedition:	Expedition 43
ISS Crew:	Terry Wayne Virts (NASA), ISS-CDR
	Samantha Cristoforetti (ESA), ISS-Flight Engineer 1
	Anton Nikolayevich Shkaplerov (RKA), ISS-Flight Engineer 2
	Gennadi Ivanovich Padalka (RKA), ISS-Flight Engineer 3
	Mikhail Borisovich Korniyenko (RKA), ISS-Flight Engineer 4
	Scott Joseph Kelly (NASA), ISS-Flight Engineer 5

The ISS Story So Far

Soyuz TMA-14M launched on September 25. The port solar array failed to deploy automatically. The impact experienced during docking caused it to "pop out" and there was no risk to the crew or mission.

Samantha Cristoforetti

Samantha Cristoforetti was born in Milan, Italy, on April 26, 1977. She studied at the Liceo Scientifico in Trento, Italy, until 1996, including a year as an exchange student in the United States. She earned a master's degree in mechanical engineering from the Technische Universität Munich, Germany, in 2001. Her specialization was aerospace propulsion and lightweight structures. Her thesis was on solid rocket propellants. She also studied at the Ecole Nationale Supérieure de l'Aéronautique et de l'Espace in Toulouse, France, and at the Mendeleev University of Chemical Technologies in Moscow, Russia.

She joined the Italian Air Force in 2001, graduating from the Air Force Academy in 2005, during which time she earned a bachelor's degree in aeronautical sciences at the University of Naples Federico II, Italy. After earning her wings at the Euro-NATO Joint Jet Pilot Training program at Sheppard Air Force Base in the United States in 2006, she joined the 51° Stromo (Bomber Wing) at Istrana, Italy, flying the AMX ground attack fighter jet. She was the first female fighter pilot in the Italian Air Force. She is now a captain and has logged over 500 hours across six types of aircraft.

She was selected as an ESA astronaut in May 2009, joined ESA in September 2009 and completed her basic astronaut training in November 2010. Once that was complete, she trained in EVA, robotics and was certified as a Soyuz flight engineer.

Fig. 5.1. Samantha Cristoforetti (ESA)

She was awarded the following honors:

- Commander of the Order of Merit of the Italian Republic, 6 March 2013
- Knight Grand Cross of the Order of Merit of the Italian Republic, 16 July 2015

The Futura Mission

Mission Patches

The mission name Futura was chosen after a call from Cristoforetti for a new mission name to be inspired by the words "research, discovery, science, technology, exploration, wonder, adventure, travel, excellence, teamwork, humanity, enthusiasm, dreams and nutrition." She said "A big thank you to all those who sent their ideas for the name of the mission that will take me to the International Space Station in less than a year now. It was not easy to choose from more than a thousand proposals. I shared some words that are dear to me, and I think 'Futura' brings them all together with plenty of positive momentum towards the future."

The patch design was by 31 years old Valerio Papeti from Turin, Italy, who won the competition. It displays a stylized orbit of the ISS passing Earth – symbolizing the connection between our planet and the orbital outpost. Valerio added the element of a sunrise from space, describing it as the most beautiful image he has seen from space. The sunrise also symbolizes the future of new discoveries and new horizons for Italy and humankind. The three colors of the national flag reflect Italy's involvement in the mission.

The Soyuz TMA-15M was designed by commander Anton Shkaplerov and cosmonaut Andrei Babkin, with graphics work by Riccardo Rossi. It is explained that "It is based on the shape and features of an attitude indicator, a fundamental element on an aircraft instrument panel. This represents the pilot profession, which is common to all three crewmembers. The outline of a Soyuz spaceship and its golden solar array panels, representing the horizontal lines of the indicator, are integrated with the instrument's pitch and bank angle scales. The depicted attitude of the spaceship corresponds to a bank angle of 15° (the serial number of this Soyuz TMA) and a pitch angle of 51° (the orbit inclination)". The Soyuz is flying above Earth heading for the ISS with the rising Sun a symbol of cognition and renewal. A shadow on Earth's surface is in the shape of an aircraft combining elements of a MiG-29, an F-16 and an AMX. These are Russian, U. S. and Italian aircraft flown by each of the three crew members in their respective air force careers. According to spacepatches. info "The three more prominent stars, next to the constellations Auriga and Cassiopeia, represent the fulfillment of a spaceflight by the cosmonaut and the two astronauts of this expedition. The overall number of stars corresponds to the last two figures of the launch year (2014) and, including the Sun, of the return year (2015)".

Fig. 5.2. Futura mission patch (ESA)

Fig. 5.3. Soyuz TMA-15M mission patch (ESA)

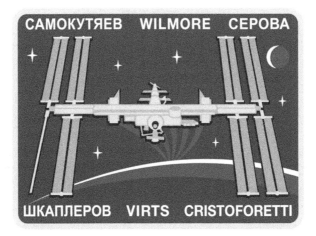

Fig. 5.4. Expedition 42 mission patch (ESA)

Fig. 5.5. Expedition 43 mission patch (ESA)

NASA explains that the Expedition 42 patch shows the "International Space Station orbiting Planet Earth with its solar array wings (SAW) spread wide. Facing the Sun with the lower left outboard solar array feathered, the left array portrays a prominent number '4' and the fully deployed arrays on the right form the Roman numeral version of '2,' which signifies the two increment crews that, together, comprise the six-member international Expedition 42 crew. The crew and all supporting personnel around the world are also represented by the six stars adorning the sky around the complex".

Likewise, the NASA description of the Expedition 43 patch explains that it was designed by Matt Lehman, assisted by Brandon Heath, working with astronaut Terry Virts. "The hexagon (six-sided) shape of the Expedition 43 patch represents the six crew members living and working on board the orbital outpost. The International Space Station is portrayed in orbit around Earth, representing the multinational partnership that has constructed, developed, and continues to operate the ISS for the benefit of all humankind. The sunrise marks the beginning of a new day, reflecting the fact that we are at the dawn of our history as a spacefaring species. The Moon and planets represent future exploration of our Solar System, for which the ISS is a steppingstone. Finally, the five stars honor the five crews who have lost their lives during the pursuit of human spaceflight: *Apollo 1, Soyuz 1, Soyuz 11,* STS-51L *Challenger* and STS-107 *Columbia*".

Fig. 5.6. The Soyuz TMA-15M crew with Samantha Cristoforetti on right (www.spacefacts.de)

Fig. 5.7. The Expedition 42 crew with Samantha Cristoforetti on right (ESA)

Fig. 5.8. The Expedition 43 crew with Samantha Cristoforetti third from right (NASA)

Timeline

Expedition 42, Week Ending November 23, 2014

The Soyuz TMA-15M launched from the Baikonur Cosmodrome in Kazakhstan to the International Space Station at 22:01 CET on November 23 carrying Terry Virts of NASA, Anton Shkaplerov of the Russian Federal Space Agency (Roscosmos) and Samantha Cristoforetti of the European Space Agency.

As was now standard with Soyuz, the astronauts reached their destination just 5 hours and 48 minutes after liftoff and four orbits. They docked 03:49 CET, November 24 and the hatch was opened at 06:00 CET. They were welcomed aboard by NASA station commander Barry Wilmore and Roscosmos cosmonauts Yelena Serova and Alexander Samokutyaev.

Expedition 42, Week Ending November 30, 2014

On November 25, after being reduced to three after the departure of Soyuz TMA-13M, the ISS was now fully crewed with six on board. This was a rest day for the crew, but November 26 they started to transfer cargo from the new Soyuz, conduct science and work on maintenance. Cristoforetti worked in the ESA Columbus laboratory module getting the European Physiology Module ready for upcoming installation work. Virts worked in the JAXA Kibō laboratory, getting a small satellite deployer ready for installation. Shkaplerov spent time on crew orientation and cargo transfers. Station commander Wilmore reviewed the new 3-D printer payload. Samokutyaev was conducting diagnostic work inside the Zarya cargo module. Serova worked on various maintenance tasks throughout the Russian segment.

Wilmore and Virts set up the Cyclops nanosatellite deployer in the Kibō laboratory on November 26. Wilmore maintained the science freezer and Virts worked on the Aniso Tubule botany study and measured air velocity in Kibō. Cristoforetti set up equipment for the Blind and Imagined experiment. This study measures visual and sensory changes in crew members on long-duration expeditions. The Russian crewmembers worked on Russian science experiments including the study of the cardiovascular system, radiation exposure in the station and plasma research. The NASA astronauts celebrated Thanksgiving on November 27 with reduced workload and shared their holiday meal with the rest of the crew.

Expedition 42, Week Ending December 7, 2014

On December 1 Wilmore and Virts conducted intricate maintenance on the Carbon Dioxide Removal Assembly, a device that removes carbon dioxide from the International Space Station's atmosphere. Samokutyaev spent the morning unloading cargo from the Progress 57, and in the afternoon, he worked maintenance in the Russian segment of the orbital laboratory. Serova updated antivirus software on laptop computers and cleaned fans and filters. She also worked on a variety of science experiments including studying blood circulation in microgravity and advanced space photography techniques.

The next day Wilmore removed and stowed a printed test object, or coupon, from the new 3D printer located in the Destiny laboratory's Microgravity Science Glovebox. Cristoforetti started the 3D print job earlier in the day. Samokutyaev and Serova partnered up on routine communications maintenance work. Shkaplerov worked on a chemistry experiment designed to educate Russian students.

Samantha Cristoforetti cleaned the BioLab on December 3. This facility in the ESA Columbus module allows the observation of microorganisms, plants and invertebrates and their adaptation to microgravity. The cosmonauts in the station's Russian segment gathered in Japan's Kibō laboratory module to record a televised event in between their regularly scheduled duties.

Expedition 42, Week Ending December 14, 2014

Cristoforetti and Wilmore reviewed procedures to replace a fan pump separator on a U. S. spacesuit on December 8. Earlier, Wilmore and Virts carried out the Body Measures experiment that studies changes to a crew member's body shape while living in microgravity. Shkaplerov and Serova got together for a chemistry education experiment during the morning. Later, Serova joined Samokutyaev for Russian maintenance work.

On December 9 Cristoforetti and Wilmore continued to work on a fan pump separator on a U. S. spacesuit. Cristoforetti started her day with medical science collecting saliva and urine samples for stowage in a science freezer. Virts worked on a variety of science projects, including updating ultrasound scanner software, checking a botany experiment and participating in an eye exam. Samokutyaev and Shkaplerov started their day in the Zarya cargo module installing an overlay sheet on interior panels. Serova conducted a photographic inspection on windows in the station's Russian segment.

By December 11 Wilmore had completed the spacesuit maintenance work and began testing the spacesuit to return it to service. Virts and Cristoforetti started their day on medical science and a periodical fitness check. Cristoforetti cleaned up and stowed the spacesuit hardware and tools. Wilmore, Cristoforetti and Virts then reviewed activities planned for the following week's SpaceX Dragon launch. Virts went on to open the Combustion Integrated Rack for fuel gear replacement work. Shkaplerov and Samokutyaev were back at work inside the Zarya cargo module installing overlay sheets on interior panels and disinfecting them. Serova was studying radiation in the station and the Sun's influence on Earth's magnetic field. Samokutyaev also joined Serova for observation of the cardiovascular system while working out on an exercise bike. Shkaplerov worked throughout the day studying chemical reactions in Earth's upper atmosphere.

Wilmore and Cristoforetti trained for the capture of the SpaceX Dragon. Wilmore scrubbed the cooling loops on a U.S. spacesuit after replacing its fan pump separator the day before. Virts sampled and tested the water conductivity inside the spacesuit. Virts, Cristoforetti and Wilmore also started their day with medical studies including a periodic fitness test and ultrasound scans of the arteries. Samokutyaev and Serova worked on an experiment that studies micrometeoroid detection techniques. Shkaplerov had an Earth photography session observing the effects of natural and manmade disasters. Virts installed a centerline berthing camera to support the mating of Dragon to the Harmony node.

Expedition 42, Week Ending December 21, 2014

Cristoforetti and Serova started December 16 conducting a test run of SPHERES. Cristoforetti checked the nitrogen pressure of science freezers in the afternoon, and then joined station commander Wilmore for a practice session with the Canadarm2 before the upcoming SpaceX Dragon mission. Wilmore began his morning with some 3D printing work before moving on to the Advanced Colloids Experiment Microscopy-3 fluids study. Virts set up the Microgravity Science Glovebox installing hardware for an experiment that studies the risk of infectious disease on long-term space missions. Samokutyaev and Shkaplerov were back at work inside the Zarya cargo module installing protective sheets on interior surfaces.

Virts turned on the Robonaut 2 on December 17 so that payload controllers from the Marshall Space Flight Center in Alabama could power up its new legs. Cristoforetti downloaded SPHERES data, demonstrating how the small free-floating satellites can build 3D maps of objects and interact and navigate using those 3D models. She also joined Wilmore removing a small satellite deployer, CYCLOPS, from the Kibō laboratory for troubleshooting. Afterward, Wilmore conducted a vision test and set up a multipurpose experiment platform in Kibō. Samokutyaev and Shkaplerov installed more overlay sheets inside the Zarya cargo module. Serova conducted a photographic inspection of the interior panels of the Zvezda service module.

NASA and SpaceX announced, on December 18, that the next Dragon would be postponed until after January 6, 2015. This was to investigate issues that arose from the static fire test of the Falcon 9 rocket on December 16. Virts worked on the Sabatier system, which produces water on the station, then joined Wilmore and Cristoforetti for a series of eye exams.

Expedition 42, Week Ending December 28, 2014

On December 22, Wilmore operated the Binary Colloidal Alloy Test before joining Virts in a call CBS Morning News and WBAL Radio in Baltimore, Maryland. Virts then gathered cargo to be returned to Earth on the Dragon capsule due to arrive at the station. He also packed rubbish in the ATV, which burns up on re-entry, unlike the Dragon which can safely return samples to Earth. Cristoforetti spoke with Giorgio Napolitano, the president of Italy and later she collected biological samples for stowage in a science freezer and worked inside the Materials Science Laboratory.

Expedition 42, Week Ending January 4, 2015

Wilmore celebrated his 52nd birthday aboard the station on December 29. The next day, Wilmore and Cristoforetti explained their Christmas and New Year's Eve plans with CBS News and the BBC. Cristoforetti installed ocular coherence tomography equipment, which records 3-D images of the retina and the interior of the eyes. Wilmore carried out eye tests on Cristoforetti and Virts. Wilmore then set up the ESA Haptics-1 experiment, which investigates how gaming technology can be used to remotely control robots from space. Serova, Samoukutyaev and Shkaplerov maintained the life support systems and checked for air leaks.

The crew orbiting Earth on the ISS got to experience New Year's Eve 16 times as they circled the globe at 17,500 miles an hour. They celebrated with fruit juice when the station officially experienced the start of 2015. That is 7 p.m. EST, or midnight by the Coordinated Universal Time (UTC) or Greenwich Mean Time (GMT).

Expedition 42, Week Ending January 11, 2015

Virts and Cristoforetti took turns on exercise cycle for fitness evaluations on January 5. Virts then replaced a dosing pump in the Waste and Hygiene Compartment. Cristoforetti took part in an experiment that explored the possibility of using plants to produce food and oxygen on the station. Shkaplerov, Samokutyaev and Serova investigated micrometeoroid impacts, carried out maintenance and photographed windows for a contamination inspection.

The launch of the SpaceX Dragon on the Falcon 9 rocket was aborted 1 minutes, 21 s before the intended launch time on January 6. A thrust vector control actuator for the Falcon 9's second stage failure was detected.

Cristoforetti participated in a study observing the aging of skin and tested an X-ray device that measures bone density in space. Virts opened the Fluids Integrated Rack to prepare samples for the Advanced Colloids Experiment-Microscopy-3 study. Serova deployed dosimeters for a radiation detection study and downloaded data collected from an earthquake experiment. Samokutyaev took photographs and recorded video documenting life on the station before an afternoon of maintenance in the Russian segment. Shkaplerov disinfected the area behind panels in the Zvezda service module. The crew had the day off on January 7 to celebrate the Russian Christmas holiday.

The SpaceX Dragon was launched from Cape Canaveral Air Force Station, Florida on January 10. It was berthed to the Harmony module of the ISS 8:54 a.m. EST on January 12. Cristoforetti Wilmore controlled the Canadarm2 robot arm.

Expedition 42, Week Ending January 18, 2015

The crew opened the hatches to the Dragon at 3:23 a.m. EST, January 13 and the crew began unpacking cargo.

At around 3 a.m. CET, January 14, Wilmore, Virts and Cristoforetti were instructed by Mission Control to wear breathing masks, to move to the Russian segment and to close the hatches to the USOS. They had received an environmental systems software alarm that monitored the atmosphere. The protection software automatically shut down one of two redundant cooling loops (Thermal Control System Loop B). Ground controllers were worried that an alarm indicated an ammonia leak. After studying sensor data, it was determined that there was no leak. The alarm was caused by a transient error message in one of the station's computer relay systems or multiplexer-demultiplexer. After the all-clear the crew re-entered the station's U. S. segment at 2:05 p.m. CET wearing protective masks, until Virts and Cristoforetti sampled the cabin atmosphere and reported no ammonia detected.

Expedition 42, Week Ending January 25, 2015

Wilmore reported an abnormally loud fan pump while scrubbing cooling loops on a U. S. spacesuit on January 20. Also reported to mission control was a bad odor in the ATV. The hatches were closed and European controllers performed waste tank leak checks and didn't find any leak indications. Samokutyaev was set to re-enter the vehicle wearing a respirator mask to investigate. It turned out to be badly sealed trash.

On January 21, Wilmore conducted botany research and harvested plants grown for the Advanced Plant Experiments-03-1 (APEX-03-1). The cress plants are photographed and preserved in a science freezer for future testing. Virts processed samples for the Coarsening in Solid Liquid Mixtures-2 (CSLM-2) experiment, which studies the processes that occur in materials for consumer and industrial products. He later unpacked more gear from inside Dragon.

Ground controllers overnight used the Canadarm2, with its Dextre robotic hand to remove the CATS experiment from the SpaceX Dragon exposed platform. They handed it off to the Japanese robotic arm to mount on the Kibō external platform. CATS, or Cloud-Aerosol Transport System, will collect data on the pollution, dust, smoke, aerosols and other particulates in Earth's atmosphere.

Wilmore replaced hardware that stores and delivers fuel, including igniter tips on the Combustion Integrated Rack (CIR), in the Destiny laboratory. Virts and Cristoforetti, with the help of doctors on the ground, participated in more eye checks as part of the Ocular Health study. Virts also checked samples and transferred data collected for the Coarsening in Solid Liquid Mixtures-2 (CSLM-2) experiment.

Expedition 42, Week Ending February 1, 2015

Virts and Wilmore prepared a pair of U. S. spacesuits on January 27, for a set of February spacewalks. They recharged suit batteries and checked out fans and other suit components. Virts also joined Cristoforetti transferring cargo to and from the SpaceX Dragon.

Virts unpacked the Robonaut in Destiny and tested its mobility using its legs for the first time.

Expedition 42, Week Ending February 8, 2015

Wilmore prepared heater cables on February 2, for installation on a future spacewalk. Virts processed samples for a materials science experiment and removed hardware from the Commercial Generic Bioprocessing Apparatus.

The next day, Wilmore continued to load the ATV-5 with trash. Samokutyaev and Shkaplerov trained with the telerobotically operated rendezvous system

(TORU) in anticipation of the arrival of Progress 58. Wilmore also harvested plants for the APEX-03 botany experiment. Cristoforetti looked at roundworms for the Epigenetics study that studied whether new cell generations adapt to weightlessness.

On February 4, European flight controllers investigated a signal indicative of a failure in a power chain that provides battery power to the ATV-5. Three other power chains were operating normally inside the ATV-5, so normal operations continued.

Samokutyaev and Shkaplerov set up equipment in the Zvezda Service Module to monitor the departure of the ATV. Wilmore and Virts checked a U. S. spacesuit and attempted to restart its fan motor after it failed. Another spacesuit with a similar issue was packed inside the SpaceX Dragon, to be investigated on Earth later.

Expedition 42, Week Ending February 15, 2015

The cosmonauts worked in the Russian segment, on February 9, studying how to find micro-meteoroid holes in the station surfaces. They examined charged macroparticles inside a magnetic trap and they also explored crew training methods using interactive 3D manuals, or virtual manuals. Cristoforetti installed cameras in the ATV that would capture the re-entry disintegration.

Wilmore and Virts disconnected cables and depressurized the vestibule between Harmony and Dragon. Cristoforetti, with help from Virts, controlled the Canadarm2 as she released Dragon at 2:10 p.m. EST. The capsule was moved outside the vicinity of the space station in preparation for its return trip to Earth. The Russian crew stowed rubbish into the Progress 57. Dragon conducted its deorbit burn on time at 5:49 p.m. CET and splashed down in the Pacific Ocean at about 7:44 p.m. EST 259 miles southwest of Long Beach, California.

The station crew also focused on spacewalk preparations and microgravity science, the primary mission of the orbital laboratory, to benefit life on Earth as well as future space crews. Ground doctors assisted Wilmore and Cristoforetti during eye exams. Cosmonauts Alexander Samokutyaev and Elena Serova studied bioelectric cardiac activity as well as methods to locate punctures caused by micrometeoroids on the station's surface.

The fifth and last ATV undocked from the station's aft port of the Zvezda service module at 8:42 a.m. EST on January 12. The five ATVs delivered approximately 34 tons of supplies to the complex while docked to the station for a total of 776 days. The Georges Lemaitre ATV entered Earth's atmosphere and burned up over the Pacific Ocean around 12:12 p.m. CET on January 15.

Expedition 42, Week Ending February 22, 2015

The Progress 58 cargo craft launched at 6:00 a.m. EST, February 17, from the Baikonur Cosmodrome in Kazakhstan. The uncrewed spacecraft docked at

11:57 a.m. EST to the rear port of the Zvezda service module. It contained 3 tons of food, fuel, supplies and experiment hardware. The hatches were opened the following day.

Ground controllers moved the Canadarm2 and replaced a faulty Remote Power Controller Module (RPCM). The RPCM provides backup commanding capability to the port Thermal Radiator Rotating Joint.

Wilmore and Virts switched their spacesuits to battery power at 7:45 a.m. EST, February 21 to start their spacewalk. They attached power and data cables on the Pressurized Mating Adapter-2 (PMA-2) which is attached to the Harmony module. This was in preparation for the arrival of the first of two International Docking Adapters (IDA) later in the year. Boeing built the two new docking adapters to be delivered to the station on a pair of SpaceX Dragons. The IDAs are needed to dock the two new crewed spacecraft, Boeing's Crew Space Transportation CST-100 and SpaceX's Crew Dragon. They also reinstalled a debris shield. They ended their spacewalk at 2:26 p.m. EST with the depressurization of the Quest airlock. The EVA lasted 6 hours and 41 minutes. It was the first for Virts, whereas Wilmore had now has spent 13 hours and 15 minutes during two EVAs.

Expedition 42, Week Ending March 1, 2015

Ground controllers maneuvered the Canadarm2 into place to support the upcoming spacewalk on 23 February. Wilmore and Virts reviewed the procedures and prepared tools and spacesuits. Cristoforetti donned wearable equipment for the Drain Brain study that studies blood-flow from the brain to the heart in space. The Russians celebrated the Russian Defender of the Fatherland holiday with a day off.

At 6:51 a.m. EST, February 24, Wilmore and Virts started their second spacewalk during which they installed power cables on the Pressurized Mating Adapter-2, lubricated the Latching End Effecter of the Canadarm2, and prepared the Tranquility module for the station's upcoming reconfiguration in preparation for the arrival of commercial crew vehicles, scheduled for 2020. The EVA lasted 6 hours, 43 minutes and ended at 1:34 p.m. EST with the repressurization of the Quest airlock.

Virts reported seeing a small amount of water floating free in his helmet during airlock repressurization at the conclusion of the spacewalk. He had not seen water in the helmet during the spacewalk. After removing the helmet, Cristoforetti confirmed the free-floating water inside the helmet and indicated the helmet absorption pad was damp. Virts was wearing spacesuit #3005, in which water was also detected on a spacewalk in December 2013. Ground teams began a detailed assessment prior to the next spacewalk.

Progress 58 fired its engines on February 26 to raise the station's orbit by 1.3 statute miles. Wilmore and Virts checked their tools and spacesuits before their third and final U. S. spacewalk of the expedition. The following day, the Mission Management Team gave the go ahead for the third EVA. They determined that the water in Virts' suit, serial number 3005, was "sublimator water carryover, a small amount of residual water in the sublimator cooling component that can condense once the environment around the suit is repressurized following its exposure to vacuum during a spacewalk, resulting in a tiny amount of water pushing into the helmet".

Expedition 42, Week Ending March 8, 2015

Virts and Wilmore started their third spacewalk at 6:52 a.m. EST, on March 1. They installed a boom and Virts installed one of two antennas for the Common Communications for Visiting Vehicles (C2V2) system on the port side. Wilmore installed the starboard antenna. No water was detected in their helmets. They ended their spacewalk at 12:30 p.m. EST with the repressurization of the Quest airlock. The EVA lasted 5 hours, 38 minutes.

On March 4, the final pair of 16 cubesats was deployed overnight from outside the Kibō laboratory. Virts and Cristoforetti conducted airflow monitor tests, measurements and calibrations in the Quest airlock. The tests were part of the Airway Monitoring experiment that is looking for possible indicators of airway inflammation in astronauts during spaceflight. Shkaplerov tested for microbes in the Zarya cargo module and took pictures of the interior surfaces. Wilmore, Samokutyaev and Serova practiced a Soyuz descent drill ahead of their departure.

On March 6, Cristoforetti and Virts monitored their breathing in the quest airlock as air pressure was reduced in the module. The Airway Monitoring experiment measures nitric oxide breathed out by the astronauts. After conducting the experiment on Earth before launch, they were now performing it in space on the station for the first time. Eight astronauts would ultimately perform this test. Pressure was reduced by 30%, the equivalent to being at 3000-m altitude. They were the first of eight astronauts to collect data on their lungs for this experiment. Lars Karlsson, lead investigator for this experiment from the Karolinska Institute of Sweden, was hopeful that the experiment in the airlock would open up new fields of research in reduced pressure in space: "In the future, it is quite likely that drugs could be designed based on exhaled nitric oxide measurements, to find the most effective molecules to treat inflamed airways and lungs. This type of research is a first step down this road."

Expedition 42/43, Week Ending March 15, 2015

On March 10, Expedition 42 commander Barry Wilmore transferred command of the International Space Station to NASA astronaut Terry Virts in a Change of Command Ceremony. After the ceremony, Virts continued to install cables for future commercial crew vehicles which will dock at the International Docking Adapters.

At 3:34 p.m. EDT on March 11, the Soyuz hatch closed between the International Space Station and the TMA-14M spacecraft. Expedition 42 crew members Barry Wilmore, Alexander Samokutyaev and Elena Serova undocked at 6:44 p.m., at which point Expedition 43 began. Their Soyuz TMA-14M spacecraft landed in Kazakhstan at approximately 10:07 p.m. EDT.

Expedition 43, Week Ending March 22, 2015

Reduced to a crew of three – station commander Terry Virts and Flight Engineers Samantha Cristoforetti and Anton Shkaplerov – continued the station's maintenance and research tasks. Virts installed a connector cap and a power cable to prepare for the relocation of the Permanent Multipurpose Module. It would be moved from the Unity module to the Tranquility module. Cristoforetti reviewed overview materials for the Muscle Atrophy Research and Exercise System (MARES). She also set up the TripleLux-B experiment which studies cellular mechanisms in space.

Cristoforetti went back to work on MARES on March 17. Virts packed a physics experiment in the Microgravity Science Glovebox in the Destiny module. He stored the Coarsening in Solid Mixtures-4 (CSLM-4) experiment fore return to Earth on a future SpaceX Dragon capsule return. The following day, Cristoforetti completed the activation and testing of MARES. She also inspected cables and connectors on a science freezer for corrosion. Virts set up the Advanced Colloids Experiment Microscopy-3 (ACE M-3). He cleaned the Microgravity Science Glovebox and certified it for another year of use.

The Progress 58 spacecraft fired its engines on March 18 for 4 minutes, 18 seconds in preparation for the arrival of the Soyuz TMA-16M, carrying Gennady Padalka and One-Year mission crew members Scott Kelly and Mikhail Kornienko. Meanwhile, Virts replaced hardware on the Waste and Hygiene Compartment. He also participated in the Astro Palate study investigating how food affects the mood of crew members during a spaceflight. Cristoforetti began the TripleLux-B experiment in the BioLab glovebox. This investigates cellular mechanisms that cause impairment of immune functions in weightlessness.

Virts inspected windows and checked for dust build-up in vents inside the Destiny laboratory module on March 20.

Expedition 43, Week Ending March 29, 2015

Virts and Cristoforetti undertook their 120-day medical tests on March 23. Virts had a vision test for the Ocular Health study. Cristoforetti collected blood and urine samples for the Biochemical Profile and Bone and Muscle Check.

On the following day, Cristoforetti ran the TripleLux-B experiment again, to establish how cellular mechanisms cause impairment of immune functions. Shkaplerov studied the effects of Earth's magnetism on the space station and radiation exposure on a mannequin fitted with sensors.

Virts and Cristoforetti started the installation of the Robotics Refueling Mission (RRM-2) payload inside the Japanese Experiment Module airlock on March 26. The RRM-2 investigation is exploring how robotics could be used to fix satellites not designed to be serviced in orbit.

On March 25, Cristoforetti spoke to a group of students aged 8 to 12 gathered in three locations on Earth. European astronauts were at each location: Paolo Nespoli was at the Muse Science Museum in Trento, Italy; Pedro Duque was at ESA/ESAC with Universidad Politecnica de Madrid in Spain; and Franz Viehböck was at the Museum of Natural History in Vienna, Austria.

The Soyuz TMA-16M launched from the Baikonur Cosmodrome in Kazakhstan at 3:42 p.m. EDT on March 27. It carried Scott Kelly of NASA, Mikhail Kornienko and Gennady Padalka of the Russian Federal Space Agency, Roscosmos. The Soyuz docked to the Poisk zenith port of the International Space Station at 9:33 p.m. EDT, while the station was over the western coast of Colombia. They were welcomed on board by Expedition 43 Commander Terry Virts of NASA, Anton Shkaplerov of Roscosmos and Samantha Cristoforetti of ESA.

Kelly and Kornienko were starting the One-Year Mission, where the effects of long-term space missions on the human body are investigated. The scientists took advantage of the fact that Scott Kelly has an identical twin, Mark Kelly, who remained on Earth and acted as a control.[1,2]

Expedition 43, Week Ending April 5, 2015

The crew participated in eye checks for the Ocular Health study as scientists investigated how microgravity affects vision during long duration missions. The new crew members trained to prepare for a medical emergency while also familiarizing themselves with station systems.

Progress 58 fired its engines on April 2, boosting the space station's orbit by 0.8 mile. The reboost readied the station to receive the new Progress 59.

---

[1] The 'year-long' mission actually lasted 343 days.

[2] Scott and Mark Kelly were both U. S. Navy aviators who reached the rank of captain. Both were selected as part of NASA Astronaut Group 16 in 1996. Both have flown in space four times. Both have been pilot and commander of space shuttle missions. Only Scott has flown on Soyuz capsules and commanded the ISS.

Ocular Health eye checks continued among the crew. New software was loaded onto computers for the Rodent Research study. Kelly collected a saliva sample and the Russian crew unloaded equipment from Soyuz TMA-16M and packed the Progress 57.

Expedition 43, Week Ending April 12, 2015

Virts and Kelly carried out spacesuit maintenance in the Quest airlock April 6. Cristoforetti worked on a variety of botany experiments. Shkaplerov, Padalka and Kornienko, worked on science and maintenance in the Russian segment, sampling air quality. They also maintained the Kurs automated rendezvous system.

On the following day, Virts measured his blood pressure so it could be compared to readings taken on the ground as part of the BP Reg experiment. Kelly collected sweat samples after a training session for the Microbiome study and prepared for ultrasound and blood pressure testing. Cristoforetti continued the botany work using the Aniso Tubule system and collected equipment for the Rodent Research study. Kornienko mixed cell cultures in a bioreactor to investigate the dynamic forces on the station. Shkaplerov completed the 24 hours blood pressure monitoring session and then worked on the Bar experiment to try to detect air leaks by monitoring air pressure differences. Cristoforetti and Virts trained with the Canadarm2 to prepare for the arrival of the SpaceX Dragon capsule.

Expedition 43, Week Ending April 19, 2015

On the morning of April 13 the crew had a break from work as they commemorated the first spaceflight of cosmonaut Yuri Gagarin on April 12, 1961. They returned to work in the afternoon with Virts, Kelly and Cristoforetti, who were participating in a Dragon cargo conference with ground controllers.

SpaceX postponed the launch of the Falcon 9 on April 13, due to weather conditions in Florida. It finally launched on the following day at 4:10 p.m. EDT. At the time of launch, the International Space Station was traveling at an altitude of 257 miles over the Great Australian Bight, south of western Australia. The spacecraft was loaded with more than 1950 kg of supplies.

On April 14, NASA Administrator Charles Bolden spoke of the success of SpaceX. He said "Five years ago this week, President Obama toured the same SpaceX launch pad used today to send supplies, research and technology development to the ISS. Back then, SpaceX hadn't even made its first orbital flight. Today, it's making regular flights to the space station and is one of two American companies, along with the Boeing Company, that will return the ability to launch NASA astronauts to the ISS from U. S. soil and land then back in the United States. That's a lot of progress in the last 5 years, with even more to come in the next five."

Virts set up hardware inside Harmony to assist Dragon's installation after its capture. Virts and Cristoforetti practiced their robotics skills necessary to capture Dragon with the Canadarm2. They captured the Dragon at 6:55 a.m. EDT while the station was traveling 257 statute miles over the Pacific Ocean just east of Japan. Ground controllers then took over the Canadarm2 to maneuver the Dragon. It was berthed at 9:29 a.m. EDT.

Fig. 5.9. Samantha Cristoforetti "There's coffee in that nebula... ehm, I mean... in that Dragon." April 2015 (ESA) (Cristoforetti wore a Star Trek Voyager uniform and paraphrased Captain Kathryn Janeway, who said "There's coffee in that nebula" in the 1995 episode, *The Cloud*, to indicate a source of fuel to power the ship's coffee-making replicators. The Dragon was carrying a new coffee maker for the station.)

Virts and Kelly opened the hatches and entered the SpaceX Dragon space freighter Saturday.

Expedition 43, Week Ending April 26, 2015

The crew started the week by unloading new Rodent Research equipment that will study the effects of microgravity on biological mechanisms in mice. They also unloaded the two new POLAR science freezers which will store science samples at minus 80 °C.

While the Dragon was being unloaded, the Progress 57 space freighter was being loaded with rubbish to burn up on re-entry.

Kelly and Virts started working on the Rodent Research facility immediately. Throughout the week, the crew conducted a variety of experiments, including the Myco experiment and the Interactions study. The formeranalyzes nose, throat and skin samples and examines how microorganisms affect allergies. The latter investigates crews from various countries work together. The usual blood pressure tests and eye checks were conducted.

The Dragon delivered a special item, the 'ISSpresso' machine to make fresh coffee for the crew. Italian coffee fan Cristoforetti was particularly pleased to see it working on the station.

The Progress 57 cargo spacecraft separated from the International Space Station at 2:41 a.m. EDT while the spacecraft were flying 257 miles above northwestern China.

Expedition 43, Week Ending May 3, 2015

The Expedition 43 crew conducted the Sprint study where they used high intensity, low volume exercise training to minimize loss of muscle, bone, and cardiovascular function in the crew. Virts maintained the Japanese airlock and got it ready for the Robotics Refueling Mission-2 (RRM-2) operations.

Progress 59 uncrewed cargo ship launched at 3:09 a.m. EDT on April 28 from the Baikonur Cosmodrome in Kazakhstan. A problem occurred where the Kurs rendezvous system did not initialize as expected. Russian ground controllers failed to gain control and fix the problem remotely. The planned docking was postponed.

By April 29, Russian controllers had abandoned hope of communicating with and controlling the Progress craft. They notified the crew. Both the Russian and US segments of the station continued to operate normally and were adequately supplied well beyond the next planned resupply flight. They began to calculate to identify the most likely period for Progress 59's entry back into Earth's atmosphere. It was predicted that the spacecraft would orbit for approximately 2 weeks before deorbiting. The U. S. Air Force Joint Functional Component Command for Space's Joint Space Operations Center was also tracking Progress, performing conjunction analysis, and providing warning of any potential collisions in space to ensure spaceflight safety. It was determined that the break up and re-entry of the Progress posed no threat to the ISS crew.

Expedition 43, Week Ending May 10, 2015

On May 4 Virts and Kelly prepared for the replacement of a Carbon Dioxide Removal Assemblies (CDRA) The CDRA removes carbon dioxide from the air. Cristoforetti conducted the Skin-B experiment to investigate accelerated skin-aging in space. She also photographed the Moon.

Controllers in Houston remotely operated the Canadarm2 to install new task boards on the Robotic Refuelling Mission-2 (RRM-2).

The following day, Cristoforetti participated in two experiments, the Triplelux-A experiment and the Osteo-4 study. The former investigates immune suppression in space while the Osteo-4 study looks at the effects of weightlessness on bone cells. She also helped Shkaplerov in a Soyuz descent training exercise as they prepare to return to Earth. On May 6, Scott Kelly spoke with the "Today Show" and his twin brother Mark Kelly. Cristoforetti and Shkaplerov also checked out the Sokol launch and entry suits that they, along with Virts, will wear when they return to Earth.

On May 7, the Progress 59 cargo craft that failed to dock with the ISS re-entered Earth's atmosphere at 10:04 p.m. EDT over the central Pacific Ocean. The loss of the spacecraft posed no threat to the ISS crew.[3]

The crew had a half day on May 8. Virts and Kelly completed their work on the Carbon Dioxide Removal Assemblies (CDRA). They reconnected power, data and fluid lines to the unit and ground controllers reinitiallised it. Cristoforetti focused on the Skin-B study and the Triplelux-A experiment.

Expedition 43, Week Ending May 17, 2015

Kelly and Virts continued working on the Rodent Research experiment this week. They examined how mice changed in microgravity. Cristoforetti configured a Microgravity Experiment Research Locker Incubators (MERLIN) for return on Dragon. She also loaded cargo for return to Earth into the Dragon on May 11.

On May 12, it was announced that the return to Earth for Virts, Cristoforetti and Shkaplerov would be postponed to early June. It was originally scheduled for late May. After the loss of the Progress vehicle, Roscosmos requested time to further investigate the cause before the crew fly in a very similar Soyuz craft.

The crew continued with their assigned tasks. Kelly examined how the fine motor skills of an astronaut change on long duration space missions. He also prac-ticed with the Robotics Refuelling Mission. Cristoforetti worked on the Rodent Research experiment during the afternoon. Virts continued to load cargo into the Dragon capsule. Shkaplerov, Padalka and Kornienko cleaned filters, replaced smoke sensors and continued to sample the atmosphere for microbes in the Russian segment as part of ongoing maintenance.

Virts replaced a fan on the Carbon Dioxide Removal Assembly on May 14. Cristoforetti prepared for the relocation of the Permanent Multipurpose Module by rerouting cables and pipework. Shkaplerov photographed the condition of the

---

[3] It was later determined that a new propellant tank design caused a resonant vibration, resulting in a hammer effect shockwave. The tank ruptured on separation and blasted the Progress into a higher orbit and set it spinning.

Zvezda service module windows. Padalka and Kornienko studied acoustic techniques finding micrometeoroid impacts quickly.

The following day, Kelly and Kornienko participated in the Fine Motor Skills experiment. This analyzes how the crew adapt to weightlessness and how they recover when the mission is complete. Virts measured airborne and surface microbes for the Microbial Observatory-1 study while Cristoforetti examined fluids and gases in the Kibō laboratory. Padalka set up video cameras and maintained the magnetometers in the Russian segment. Shkaplerov maintained the docking ports in the Russian segment and took photographs of the Pirs and Poisk portholes.

On Sunday, the engines of the Progress 58 cargo craft were used to boost the station. The burn lasted 32 minutes and 3 seconds raised the station's altitude by 1.2 statute miles at apogee and 3 miles at perigee and left the station in an orbit of 252.2 × 247.1 statute miles.

Expedition 43, Week Ending May 24, 2015

On May 19 the crew prepared for the departure of the SpaceX Dragon. They reviewed and tested the Commercial Orbital Transport Services (COTS) UHF Communication Unit (CUCU) used to communicate with the Dragon capsule. They also stowed the last items of cargo into the ship. Kornienko took samples for the Fluid Shifts vision experiment and Cristoforetti also scanned Virts' eyes as part of the Ocular Health experiment.

The next day, Kelly and Virts moved samples from MELFI freezers to the Dragon, including results of the Cell Shape and Expression, CASIS PCG-3, Nematode Muscles experiments. Dragon undocked from the Earth-facing side of the station's Harmony module on May 21 and was released by the Canadarm2 at 7:04 a.m. EDT. The Dragon finally splashed down in the Pacific Ocean at 12:42 p.m. EDT, about 155 miles southwest of Long Beach, California.

Expedition 43, Week Ending May 31, 2015

Station commander Virts and Kelly closed the hatches of the Permanent Multipurpose Module (PMM) before it is relocated on May 26. Cristoforetti moved some experiment to the external platform of the Kibō laboratory using the Small Fine Arm robot and she recorded her actions. The Russian crewmembers continued their analysis of how sound waves could help find micrometeoroid impacts.

On May 27 the Permanent Multipurpose Module (PMM) was detached from a berthing mechanism on the Earth-facing port of the Unity module at 5:50 a.m. EDT by robotics flight controllers at NASA Mission Control, Houston, and at the Canadian Space Agency (CSA) centre in St. Hubert, Quebec, Canada. The 11-ton PMM was used for storage and was maneuvered to an installation position at the forward port of the Tranquility module. This freed the Unity port as a second berthing location for U. S. commercial cargo spacecraft. That means a total of four ports for U. S. vehicles. The PMM was successfully relocated from the Unity module to the Tranquility module at 9:08 a.m. EDT.

Leonardo was built and designed by Italy's ASI space agency to transport cargo and equipment to the space station inside NASA's space shuttle. Modified to improve its shielding and visibility to visiting craft, it was attached permanently to the station in 2011. In exchange for supplying *Leonardo*, NASA agreed that ASI would send astronauts to the station, such as Cristoforetti.

Meanwhile, Kelly and Padalka worked on the Fluid Shifts study while Kornienko worked on the Morze study which examines how fluid changes in space can affect a crew member's immunology and blood pressure.

Expedition 43, Week Ending June 7, 2015

Kelly gathered hardware on June 1 for the eye-health Fluid Shifts experiment. Virts worked on the Multi-Purpose Small Payload Rack (MSPR) and Thermal Container so that ground controllers can collect data for the Cell Mechanosensing-3 experiment. Cristoforetti took samples for the Microbiome experiment to see how the microbes in the human body (microbiome) adjust to spaceflight. Virts and Cristoforetti collected tools and equipment in the Quest airlock in order to maintain the U. S. spacesuits.

American astronauts Virts and Kelly commemorated 50 years of operations in Houston's Mission Control Center. On June 3, 1965, Mission Control Center in Houston began operations when Gemini 4 lifted off, carrying James McDivitt and Ed White. The mission saw the first American spacewalk.

Padalka studied plasma crystals that could benefit future spacecraft design. Cristoforetti continued her work with the Rodent Research experiment. Kornienko was back at work on the Fluid Shifts study.

Roscosmos announced on June 5 that Expedition 43 crew members Terry Virts, Anton Shkaplerov and Samantha Cristoforetti would return to Earth on June 11 at 9:43 a.m. EDT.

Expedition 43, Week Ending June 14, 2015

On June 8 flight controllers initiated a pre-determined avoidance manoeuvre (PDAM) to avoid a possible collision with a fragment of a spent Minotaur rocket body launched in 2013. They fired the thrusters on the Progress 58 for 5 minutes, 22 seconds at 2:58 p.m. CDT to slightly increase the station's orbit. The debris was projected to pass within 3 statute miles of the station. The maneuver raised the station's altitude by 106 feet at apogee and 0.7 mile at perigee, resulting in an ISS orbit of 254.0 × 244.8 statute miles.

The next day, Virts, Shkaplerov and Cristoforetti packed their Soyuz TMA-15M, ready for undocking and return to Earth.

At 10:27 a.m. CET, the thrusters of Soyuz TMA-15M unintentionally fired during routine testing of communications systems. This resulted in a change in the station's orientation. There was no threat to the crew or the station itself, and the issue had no impact on a nominal return to Earth of the Soyuz TMA-15M on Thursday.

Commander Terry Virts handed over command of the International Space Station to cosmonaut Gennady Padalka on June 10.[4]

At 3:04 a.m. EDT on June 11, the hatches closed between the International Space Station and the TMA-15M spacecraft. Expedition 43 crew members Terry Virts of NASA, Samantha Cristoforetti of ESA and Anton Shkaplerov of Roscosmos undocked at 6:20 a.m. and landed in Kazakhstan at 9:44 a.m. EDT.

Postscript

At the time of writing, Futura was Samantha Cristoforetti's last spaceflight. ESA's Director General, Jan Wörner, has stated that it is ESA's intention that all astronauts in the class of 2009 will fly twice by 2024.

Cristoforetti is in a relationship with ESA astronaut trainer, Lionel Ferra, and they have a daughter, who was born in 2016. Cristoforetti enjoys hiking scuba diving, yoga and learning foreign languages. Having mastered Russian as part of her training, she is currently learning Chinese.

She has technical and management duties at the European Astronaut Center, which include serving on technical evaluation boards for exploration-related projects. For several years she led the Spaceship EAC initiative, a student-centered team working on the technological challenges of future missions to the Moon. She is now crew representative for ESA in the Lunar Orbital Platform Gateway project.

She is also participating in a working group with Chinese counterparts defining and implementing cooperation in the field of astronaut operations. In 2017, together with fellow ESA astronaut Matthias Maurer, she participated in a sea survival exercise organized by the Astronaut Center of China in the Yellow Sea. This was the first joint training of Chinese and non-Chinese astronauts in China.

In June 2019, Cristoforetti commanded the NEEMO 23 underwater expedition.

---

[4] This would be Padalka's fifth time commanding a space station, having commanded Mir in 1998 and the ISS on four occasions (2004, 2009, 2012 and 2015). He also holds the record for most time in space at 879 days.

Fig. 5.10. Samantha Cristoforetti, top right, outside the NEEMO Aquarius underwater habitat, June 2019 (ESA)

At the time of writing, Cristoforetti has accumulated 199 days, 16 hours and 42 minutes in space.[5]

---

[5] Cristoforetti's 199 days was the longest duration single spaceflight by a woman until Peggy Whitson spent 289 days in space during Expeditions 50/51. Whitson holds the record for the most cumulative time in space for any American or non-Russian, male or female, at 665 days, 22 hours and 22 minutes.

6

Iriss

Mission

ESA Mission Name:	Iriss
Astronaut:	Andreas Enevold Mogensen
Mission Duration:	9 days, 20 hours, 13 minutes
Mission Sponsors:	ESA
ISS Milestones:	ISS 44S, 81st crewed mission to the ISS

Launch

Launch Date/Time:	September 2, 2015, 04:37 UTC
Launch Site:	Pad 1, Baikonur Cosmodrome, Kazakhstan
Launch Vehicle:	Soyuz TMA
Launch Mission:	Soyuz TMA-18M
Launch Vehicle Crew:	Sergei Aleksandrovich Volkov (RKA), CDR
	Andreas Enevold Mogensen (ESA), Flight Engineer
	Aydyn Akanovich Aimbetov (KazCosmos) Flight Engineer

Docking

Soyuz TMA-18M	
Docking Date/Time:	September 4, 2015, 07:39 UTC
Docking Port:	Poisk zenith
Soyuz TMA-16M	
Undocking Date/Time:	September 11, 2015, 21:29 UTC
Docking Port:	Zvezda aft

Landing

Landing Date/Time:	September 12, 2015, 00:51 UTC
Landing Site:	near Dzhezkazgan, Kazakhstan
Landing Vehicle:	Soyuz TMA
Landing Mission:	Soyuz TMA-16M
Landing Vehicle Crew:	Gennadi Ivanovich Padalka (RKA), CDR
	Andreas Enevold Mogensen (ESA), Flight Engineer
	Aydyn Akanovich Aimbetov (KazCosmos) Flight Engineer

© Springer Nature Switzerland AG 2020
J. O'Sullivan, *European Missions to the International Space Station*,
Springer Praxis Books, https://doi.org/10.1007/978-3-030-30326-6_6

ISS Expeditions

ISS Expedition:	Expedition 44
ISS Crew:	Gennadi Ivanovich Padalka (RKA), ISS-CDR
	Mikhail Borisovich Korniyenko (RKA), ISS-Flight Engineer 1
	Scott Joseph Kelly (NASA), ISS-Flight Engineer 2
	Oleg Dmitriyevich Kononenko (RKA), ISS-Flight Engineer 3
	Kimiya Yui (JAXA), ISS-Flight Engineer 4
	Kjell Norwood Lindgren (NASA) ISS-Flight Engineer 5

The ISS Story So Far

In March 2015, Soyuz TMA-16M carried two Russians and one American to the station. After serving the usual 6 months tour, Gennadi Padalka returned home, but Mikhail Korniyenko and Scott Kelly remained in space for 340 days for a study designed to gather data relevant to a possible future mission to Mars. Kelly's twin brother and fellow astronaut, Mark, stayed on the ground and participated in comparable medical and physiological experiments as a 'control subject.'

Soyuz TMA-17M delivered crew to Expeditions 44/45, including Japanese astronaut Kimiya Yui.

Andreas Mogensen

Andrea Mogensen was born in Copenhagen, Denmark on November 2, 1976. He attended secondary school at the Copenhagen International School, graduating with an International Baccalaureate in 1995. He went on to earn a master's degree in aeronautical engineering from Imperial College London in 1999, including a semester at the Instituto Superior Tecnico in Lisbon, Portugal.

His first job as an engineer, in 2000, was working for Schlumberger Oilfield Services, as a drilling services engineer in the Republic of Congo and the Republic of Angola working on offshore oil rigs. From 2001 to 2003, he worked at Vestas Wind Systems in Ringkøbing, Denmark, as a control systems engineer in the research and development department, where he designed control systems for wind turbines. From 2004 to 2007, he was a research assistant at the Center for Space Research and a teaching assistant in the Department of Aerospace Engineering at the University of Texas at Austin in the United States. During this he studied for his doctorate in aerospace engineering at the same institution and was awarded his Ph.D. in 2007. His specialties included guidance, navigation and control of spacecraft during entry, descent and landing, mission analysis and design and trajectory optimization.

From 2007 to 2008, he worked as attitude and orbit control systems engineer for HE Space Operations, subcontracted to Airbus in Friedrichshafen, Germany, for the duration of his employment, where he worked on ESA's Swarm mission. In 2009, he was briefly a research fellow at the Surrey Space Centre at the University of Surrey in the UK, focusing on spacecraft guidance, navigation and control during entry, descent and landing for lunar missions.

Fig. 6.1. Andreas Mogensen (ESA)

He was selected as an ESA astronaut in May 2009, completed the astronaut basic training program at the European Astronaut Centre in Cologne, Germany, in November 2010. Since then he successfully gained his private pilot license and qualified for spacewalks using both the American EMU spacesuit and the Russian Orlan suit.

He participated in the 2012 ESA CAVES underground training course, alongside David Saint-Jacques of CSA, Soichi Noguchi of JAXA, Michael Finke and Andrew Feustel of NASA and Nikolai Tikhonov of Roscosmos. He was also a member of the NEEMO 17 (also called SEATEST 2) and NEEMO 19 crews in September 2013 and September 2014, respectively.

In addition to his astronaut training activities, Andreas worked on the engineering team for the ESA Lunar Lander mission at ESTEC, in the Netherlands, where he was involved in the design of the guidance, navigation and control system for precision landing.

He is married to Cecilie Beyer. He enjoys rugby, basketball, squash, scuba diving, sky diving, kite surfing, kayaking and mountaineering. Other interests include science, in particular astrophysics, exobiology, and evolution.

He is a member of the following organizations:

- the American Institute of Aeronautics and Astronautics (AIAA)
- the American Astronautical Society (AAS)
- the American Association for the Advancement of Science (AAAS)
- the Association of Space Explorers (ASE)

He was awarded the following honors:

- The Danish Royal Medal of Recompense in gold with crown and inscription, 2015
- NASA's Distinguished Public Service Medal, 2015
- Ellehammer Prize from the Society of Danish Aviation Journalists, 2015
- Genius Prize from the Society of Danish Science Journalists, 2016

Mission Patches

The mission name Iriss was chosen from over 700 suggestions received from across Europe. The winner was Filippo Magni from Italy. The name combines the Greek goddess Iris and the International Space Station ISS. Iris was the messenger of the gods of Olympus and the personification of the rainbow. As messenger, she represents the link between humanity and the cosmos, and between the heavens and Earth. The name was announced by Mogensen on April 27, 2014 at the Copenhagen at the Science Forum. Andreas said "Science has been a bridge between East and West, helping to foster peace and understanding. The name Iriss perfectly captures this aspect of the International Space Station." Filippo said "The rainbow is a universal symbol of peace and I like to think of the Space Station as a rare and outstanding example of peace and international cooperation. The figure of Iris the messenger matches the role of the astronauts who have the task of transmitting their special experience in orbit to those who remain here on Earth."

Another competition run in Denmark had 500 entrants and two designs were chosen, the mission logo and the education logo. The winning mission logo was by Poul Rasmussen he explained that "It was inspired by the Greek goddess Iris, who is often depicted with wings. The wings in the design also represent a Viking ship as used to explore the world and seek unknown horizons. The Iriss name is written in two colors to highlight that Andreas is leaving Earth for the International Space Station. Stars and planets appear above the name in stylized orbits."

The educational logo was designed by 19-year-old Louise Nielsen, who adopted the same colors as the mission logo, but the design shows a cartoon-style Andreas heading into space in a rocket.

Fig. 6.2. Iriss mission patch (ESA)

Fig. 6.3. Iriss education mission patch (ESA)

Fig. 6.4. Soyuz TMA-18M mission patch (www.spacefacts.de)

The basis for Soyuz TMA-18M patch was designed by Jorge Cartes for Expedition 44 in March 2013. An American design was chosen for that mission, so Jorge's idea was adapted as a Soyuz mission logo, at first not for any particular crew. With its pentagonal shape, it was aimed at a 2015 mission to commemorate the 40th anniversary of the Apollo-Soyuz Test Project. In May 2013, Jorge was informed that the patch would be proposed to the Soyuz TMA-16M crew, at that time expected to be commanded by Yuri Lonchakov. When Lonchakov was replaced by Gennadi Padalka in August 2013, Jorge's artwork was moved up to TMA-18M. (Luc van den Abeelen had designed the patches for Padalka's previous four flights, so he was a more logical choice for TMA-16M; proposals for TMA-17M also already existed.) Since soprano Sarah Brightman[1] was expected to be aboard Soyuz TMA-18M, Jorge was asked to add some musical symbols to the design. Sergei Volkov accepted the patch in October 2013, and it was approved by Roscosmos on April 23, 2014. The patch artwork was revealed on Spacepatches.nl on March 5, 2015. When it became apparent that Brightman would not fly, Volkov insisted that the musical notes should remain on the patch, as a more generic "music of space" theme, so only her name was replaced by that of Aimbetov.

---

[1] Soprano Sarah Brightman, who had been due to fly as a space tourist, declined the flight for family reasons, as did her backup, Japanese entrepreneur Satoshi Takamatsu.

Fig. 6.5. The Soyuz TMA-18M crew with Andreas Mogensen on right (ESA)

Timeline

Flight Day 1 – September 2, 2015

Soyuz TMA-18M was launched at 04:37:43 GMT on September 2, 2015, from Baikonur cosmodrome, Kazakhstan. It carried ESA's Andreas Mogensen, Sergei Volkov of Roscosmos and Aidyn Aimbetov of the Kazakhstan Space Agency, KazCosmos.

The mission objective was to deliver a Soyuz spacecraft to the station for the return of the One-Year Mission crew, Scott Kelly and Mikhail Kornienko. Later models of the Soyuz can remain operational parked at the station for over 6 months, so the norm is for a three-person crew to launch and land using the same Soyuz. The One-Year mission precluded this, so Soyuz TMA-18M was delivering Volkov to be part of Expeditions 45 and 46, after which he would return with Kelly and Kornienko. Mogensen and Aimbetov were to return in 10 days on Soyuz TMA-16M, commanded by Gennady Padalka.

A station debris avoidance maneuver carried out in August 2015 meant that the station was not in a position for a four-orbit fast rendezvous. For this reason, the "old school" two-day approach was taken.

When asked why the 10 days mission could not be extended by 2 days to make back the lost time, Mission Director Roland Luettgens explained:

The option was discussed by the partners that run the International Space Station, but the problem was the landing. Landing a Soyuz spacecraft is an intense affair for everyone involved, and the conditions on Earth should be as good as possible. Weather conditions, the position of the Sun in the sky and orbital parameters all play a role in the landing on the steppes of Kazakhstan. Extending the mission by two days was theoretically possible, but the conditions at the landing site were less ideal and more importantly there would be no backup dates for landing. If the weather changes last-minute astronauts in a Soyuz can wait another day before going in to land, but this would not have been possible if the Iriss mission had been extended by two days. Understandably, the International Space Station partners decided this was a risk they were not willing to take.

Flight Day 2 – September 3, 2015

The Soyuz fired its thrusters to fine-tune the orbit. Flight dynamics confirmed the Soyuz TMA-18M's new orbit was an elliptical orbit between 240 km and 293 km altitude. The burn was completed with an increase of speed, delta-v, of 1.08 m/s.

Flight Day 3 –September 4, 2015

At 05:40 GMT mission controllers in Houston transferred control of the station's attitude keeping to Moscow Mission Control in Russia. The Russians flipped the station 180° using its thrusters so the Russian segment was flying forward. Thus, the Soyuz TMA-18M's docking port was in direct sunlight during the approach and the station's solar panels were also generating power during the docking. The station's Kurs antennas were also perfectly visible throughout.

Mogensen, Volkov and Aimbetov arrived at the ISS at 07:39 GMT. They spent 2 hours preparing to open the hatch, testing for leaks and removing their Sokol suits. The hatches between the Soyuz TMA-18M spacecraft and the ISS opened at 6:15 a.m. EDT. Expedition 44 commander Gennady Padalka of Roscosmos, Scott Kelly and Kjell Lindgren of NASA, Oleg Kononenko and Mikhail Kornienko of Roscosmos and Kimiya Yui of JAXA welcomed the new crew members, marking the first time since 2013 that nine people were aboard the ISS.

Volkov with Mogensen unloaded cargo from the Soyuz. The first was the Endothelial Cells experiment that contained human blood cells. Yui took over the packages from Volkov and Mogensen and put them in Kubik's centrifuge incuba-tor. Kelly measured Andreas's height for the SkinSuit experiment. Some astro-nauts' spines have been shown to lengthen as much as 7 cm in weightlessness, which can cause pain. The Skinsuit resembles overalls that are specially designed to simulate gravitational forces from Earth to constrict the body from shoulders to feet. Andreas unpacked salad rocket seeds for an educational activity for future ESA astronaut Timothy Peake, due on the station in December.

While on the station, Mogensen ate muesli bars for the MELiSSA experiment. According to ESA "The Micro-Ecological Life Support System Alternative study aims to have a living ecosystem that will sustain astronauts indefinitely by recycling. By finely tuning how microbiological cells, chemicals, catalysts, algae, bacteria and plants interact we could process waste to deliver never-ending fresh supplies of oxygen, water and food. The Spirulina bacteria could form an important part of the MELiSSA-loop: it turns carbon dioxide into oxygen, multiplies rapidly and can also be eaten as a delicious protein-rich astronaut meal".

Flight Day 4 – September 5, 2015

Andreas used a dedicated membrane to process three bags of waste water from the station. The membrane mimics living cells. This was part of the Aquaporin Inside study. He flushed the membrane modules five times to ensure proper circulation. As the water flows from one side of the membrane to the other it leaves unwanted organic molecules behind.

After lunch he recorded himself in the Cupola observatory for students and talked to the prime minister of Denmark, Lars Løkke Rasmussen. His backup, ESA astronaut Thomas Pesquet was in Denmark and moderated the call.

Andreas was supported by the Danish Aerospace Company from their control room in Odense, Denmark, for the Aquamembrane experiment as well as for MobileHR, Interact, Space Headaches and his educational activities. Poul Knudsen, Danish Aerospace Company's USOC manager, explained "Experiments on the International Space Station require a lot of preparation. First the experiment has to be accepted by ESA, and then the hardware to support the experiment must be designed and built. The next steps are to develop test procedures, training material, train crew and perform ground tests. Usually several years passes by between experiment selection and experiment execution on ISS. The Danish Aerospace Company has been involved in the preparation of the mission of Andreas Mogensen for several years and will support Andreas every day from 8:00 to 22:00 from the control centre in Odense."

"We are ready to support Andreas Mogensen from launch to landing. We wish him all the best of luck in the world on his historic and exciting journey. He has worked hard to get to this point, and we hope he will enjoy the fantastic experience. Our team will be following him closely from ground in Odense, and we will be supporting him in many of the experiments." says Thomas A. E. Andersen, CEO of Danish Aerospace.

A Danish cubesat, AAUSAT-5, built by students at Aalborg University, was on board the station at this time, having been delivered by the Japanese HTV-5 in August 2015. It was not possible to schedule its launch from the Kibō laboratory during Mogensen's visit to the station.[2]

Cosmonaut Gennady Padalka handed command of the International Space Station to astronaut Scott Kelly.

Flight Day 5 – September 6, 2015

Using the on-board training application 3DVit, Mogensen prepared to control a robot on Earth from space using a force-feedback joystick. With Interact, ESA's Telerobotics and Haptics team aimed to validate advanced robotic control developed for future exploration programmes. The robot hand is the Interact Centaur robot at ESA's ESTEC. Mogensen transferred the 3DViT training content from the station laptop to his iPad.

Flight Day 6 – September 7, 2015

Mogensen wore the SkinSuit on September 7 and undertook a 90 minutes training session on the MobileHR wireless bike system. He used the 3D training tool called 3Dvit to setup the operation of the Centaur robot at ESTEC in the Netherlands. ~He was able to place a peg in a hole on the surface while he was orbiting in the ISS at 400 km altitude at a speed of 28,800 km/h. He remotely operated the robot to a tolerance of only 150 μm. Although his first attempt took 45 minutes, the next effort took 10 minutes. He also "touched" objects contrasted how they "felt".

He also remotely controlled a rover in the Netherlands and spoke with Danish press and his Cecilie. She asked him if this whole experience in space will affect him as a human being when he comes back to Earth. Andreas replied: "My view may change a bit, but I think I'll leave the International Space Station with a lot of amazing experiences that I will always remember."

Flight Day 7 – September 8, 2015

Mogensen completed the two-day Supvis-E activities in one day. He remotely controlled two rovers on the ground in the Netherlands and simulated the repair of a planetary base, also remotely from the station. This was enabled by the efficiency of Mogensen and the Meteron team on the ground. As a result, Iriss mission control reshuffled the timeline for flight day 7 to give him some extra free time.

While in communication with ESTEC, he announced the five winners of the national drawing competition "Draw an astronaut." He controlled the Eurobot, at

---

[2] AAUSAT-5 was launched on October 5, 2015.

ESTEC in Noordwijk, to reveal the winner. Danish Students from 4 to 10 years old entered the competition to draw an astronaut in different situations. The sketches included astronauts training at ESA, launching in the Soyuz rocket, docking at the ISS and landing back on Earth.

Mogensen and joined Padalka for a Soyuz descent drill as they prepared to head home. Meanwhile, Kelly and Lindgren tested the humanoid Robonaut. Yui checked out rodent research equipment. Volkov worked with Mogensen on the Muscle Atrophy Research Exercise System.

As a special treat Mogensen served a meal for the entire crew, including a main course, a dessert and a chocolate surprise. This was designed and prepared in advance by Michelin-star chef Thorsten Schmidt from Jutland, Denmark. The meal consisted of brisket with cabbage, vanilla custard with rhubarb and 'Space Rocks' chocolate. The food had to be ready for packing in April and had to be pre-approved by Roscosmos, ESA and NASA. Schmidt explained a surprise for the crew:

> We made some chocolates, which we ended up calling Space Rocks because they resemble meteorites. We made a personal Space Rock for each astronaut with a surprise in it. We sent the astronauts' wives and family paper pieces of 5 × 5 cm and asked them to send greetings to their men and send the greeting back to us so we could put it in the chocolate. We managed to get the letters, but only just before the deadline! You can imagine how happy an unexpected handwritten letter can make you – and not just any text but from someone who is very dear to you. I wanted to create a personal moment for each astronaut, to surprise them pleasantly.

Flight Day 8 – September 9, 2015

Mogensen and Volkov set up the Muscle Atrophy Research and Exercise System (MARES) equipment for use the following day. Mogensen was the first astronaut to use MARES in space with a subject in situ. In this case, he was the subject. This is a muscle-measurement machine to determine how muscle strength declines during long duration space missions. Samantha Cristoforetti set up the equipment during her earlier mission but did not participate. She ran the motors without sitting inside.

Yui shut down the Endothelial Cells experiment in ESA's Kubik incubator centrifuge on September 9, to prepare for the sample's return to Earth with Mogensen. He shut down the centrifuge spin and reduced the temperature from 37° C to 6° C for storage. Mogensen continued the Demes study which investigates if taste (of food etc.) is affected by microgravity. Later, he prepared the MobiPV head-mounted cameras and smartphones to examine how astronauts can have more real-time communication with ground controllers during operations and experiments.

Mogensen also tested an interactive planning viewer called Playbook. It runs from a web browser and allows a drag and drop interface for an astronaut to move scheduled tasks around with freedom. This was a planned replacement for the network-connected schedule planner called Onboard Short-Term Planner View (OSTPV). This older system required constant communication between the ground controllers and the station either on the radio or through a satellite-driven computer network. He also looked for cloud turrets as part of the Thor experiment and took 160 photos of the elusive weather phenomenon to be processed by researchers on ground.

The new commander, Kelly worked in the Japanese Kibō laboratory to set up equipment to deploy two cubesats. Lindgren spent the afternoon troubleshooting the humanoid Robonaut 2. Padalka continued ongoing Russian science studying plasma physics.

Flight Day 9 – September 10, 2015

Padalka packed equipment inside the Soyuz TMA-16M spacecraft. Kelly worked with the SPHERES. Volkov and Kornienko reviewed procedures and hardware they would use in the event of a medical emergency in space. Lindgren worked on plumbing tasks before videotaping crew activities with an IMAX camera. Yui conducted a variety of life science experiments and Kononenko worked maintenance on Russian hardware.

Flight Day 10 – September 11, 2015

Mogensen was so efficient in performing his planned tasks that he did not have activities on his last day on the station. In addition, the rest of the crew members were allowed to sleep longer on Friday in anticipation of a busy departure day. On the last day, Mogensen tested the MobiPV equipment, he completed a final Space Headaches questionnaire and he recorded video from the Japanese Kibō module. After lunch, he finished packing the Soyuz TMA-16M.

As part of the Thor program, Mogensen took images of thunderclouds over Mexico and the Caribbean. The images were sent to the Danish university DTU. The principal investigator for the Thor experiment, Torsten Neubert said:

> *The main purpose of Andreas's Thor experiment was to test the procedure, to see if the concept of imaging storms from space for scientific purposes holds. I must say, it is a real success. The images are razor-sharp, and exactly, what I had hoped to see. They show where the clouds reach high into the atmosphere pulling up water to the edge of the stratosphere. Next step is to couple the images with lightning data and cloud data from regular weather satellites.*

At 18:13 GMT Mogensen, Padalka and Aimbetov said goodbye and closed the hatch to their Soyuz spacecraft. The remaining crew prepared to photograph the departing Soyuz, a protocol in place since 2001. In 2001 the departing Progress, M-45, left behind one of the rubber O-rings stuck to the docking port. The next Progress (M1-7) could not make a hard dock as a result. From then on, all departing Russian spacecraft are photographed. The Soyuz TMA-16M undocked from the station at 5:29 p.m. EDT. Mogensen, Padalka and Aimbetov landed on September 12 at 00:51 GMT in the steppe of Kazakhstan.

Postscript

At the time of writing, Iriss was Andreas Mogensen's last spaceflight.

Andreas is currently serving as the European astronaut liaison officer to NASA's Johnson Space Center (JSC). His duties at JSC include communicating with the astronauts on board the International Space Station as capcom from Mission Control Center in Houston. Andreas was lead capcom for the SpaceX-12 and SpaceX-14 resupply missions, Ground-IV for US EVA 51 and he was Astronaut Increment Lead for Expedition 56 and 57.

He is adjunct lecturer at both the Technical University of Denmark (DTU Space) and the University of Aalborg.

Fig. 6.6. Andreas Mogensen, pictured with Queen Margrethe of Denmark and Minister of Education and Research Esben Lunde Larsen, received the Danish Royal Medal of Recompense, September 2015 (Sermitsiaq)

During his single spaceflight, Mogensen logged 9 days, 20 hours and 14 minutes in space.

7

Principia

Mission

ESA Mission Name:	Principia
Astronaut:	Timothy Nigel Peake
Mission Duration:	185 days, 22 hours, 11 minutes
Mission Sponsors:	ESA
ISS Milestones:	ISS 45S, 82nd crewed mission to the ISS

Launch

Launch Date/Time:	December 15, 2015, 11:03 UTC
Launch Site:	Pad 1, Baikonur Cosmodrome, Kazakhstan
Launch Vehicle:	Soyuz TMA
Launch Mission:	Soyuz TMA-19M
Launch Vehicle Crew:	Yuri Ivanovich Malenchenko (RKA), CDR
	Timothy Lennart Kopra (NASA), Flight Engineer
	Timothy Nigel Peake (ESA), Flight Engineer

Docking

Soyuz TMA-19M	
Docking Date/Time:	December 15, 2015, 17:33 UTC
Undocking Date/Time:	June 18, 2016, 05:52 UTC
Docking Port:	Rassvet nadir

Landing

Landing Date/Time:	June 18, 2016, 09:15 UTC
Landing Site:	near Dzhezkazgan, Kazakhstan
Landing Vehicle:	Soyuz TMA
Landing Mission:	Soyuz TMA-19M
Landing Vehicle Crew:	Yuri Ivanovich Malenchenko (RKA), CDR
	Timothy Lennart Kopra (NASA), Flight Engineer
	Timothy Nigel Peake (ESA), Flight Engineer

© Springer Nature Switzerland AG 2020   148
J. O'Sullivan, *European Missions to the International Space Station*,
Springer Praxis Books, https://doi.org/10.1007/978-3-030-30326-6_7

ISS Expeditions

ISS Expedition:	Expedition 46
ISS Crew:	Scott Joseph Kelly (NASA), ISS-CDR
	Mikhail Borisovich Korniyenko (RKA), ISS-Flight Engineer 1
	Sergei Aleksandrovich Volkov (RKA), ISS-Flight Engineer 2
	Yuri Ivanovich Malenchenko (RKA), ISS-Flight Engineer 3
	Timothy Lennart Kopra (NASA), ISS-Flight Engineer 4
	Timothy Nigel Peake (ESA), ISS-Flight Engineer 5
ISS Expedition:	Expedition 47
ISS Crew:	Timothy Lennart Kopra (NASA), ISS-CDR
	Timothy Nigel Peake (ESA), ISS-Flight Engineer 1
	Yuri Ivanovich Malenchenko (RKA), ISS-Flight Engineer 2
	Aleksei Nikolaevich Ovchinin (RKA), ISS-Flight Engineer 3
	Oleg Ivanovich Skripochka (RKA), ISS-Flight Engineer 4
	Jeffrey Nels Williams (NASA), ISS-Flight Engineer 5

The ISS Story So Far

There were no spaceflights between Soyuz TMA-18M and this mission.

Timothy Peake

Timothy Peake was born in Chichester, England, on April 7, 1972. He attended secondary school at Chichester High School for Boys in West Sussex, England, finishing in 1990. In 1992, he graduated from the Royal Military Academy Sandhurst as an officer in the British Army Air Corps. Before flying, he was attached to the Royal Green Jackets and served as platoon commander in Northern Ireland. He was awarded his Army Flying Wings in 1994 and went on to fly combat missions in the Former Republic of Yugoslavia and Afghanistan. He qualified as a helicopter instructor in 1998 at the Defense Helicopter Flying School at RAF Shawbury in Shropshire. He was promoted to Major in 2004 and attended the Empire Test Pilots' School (ETPS) at Boscombe Down, graduating in 2005 and received a Bachelor of Science degree in flight dynamics and evaluation from the University of Portsmouth in 2006.

He retired from the military in 2009 and worked as senior test pilot for AgustaWestland helicopters. He has over 3000 hours flying time on a variety of helicopters and fixed wing aircraft, including Apache, Hawk, Dakota and Mil Mi-17. He continues to fly privately.

He was selected as an ESA astronaut in May 2009 and graduated from astronaut basic training in November 2010. For the next 3 years he underwent further training and working as a communicator.

Peake participated in the first CAVES mission in 2011 along with fellow ESA astronaut recruit, Thomas Pesquet, Kanai Norishige of JAXA, Randy Bresnik of NASA and Sergei Ryzhikov of Roscosmos. He was also a member of the 16th NEEMO crew in June 2012. On this mission, his astronaut companions were Dorothy Metcalf-Lindenburger (NASA) and Kimiya Yui (JAXA).

He was a backup for the Expedition 44 crew who launched in July 2015.

Fig. 7.1. Tim Peake (ESA)

He was awarded the following honors:

- Companion of the Order of St Michael and St George (CMG), 2016
- Freedom of the city of Chichester, 2016
- Empire Test Pilot School Westland's Trophy for best rotary wing student

Peake is married to Rebecca, has two sons and enjoys climbing, caving, cross-country running and triathlon. He has written the following books:

- *Hello, Is This Planet Earth?: My View from the International Space Station*
- *Ask an Astronaut: My Guide to Life in Space*
- *The Astronaut Selection Test Book: Do You Have What It Takes for Space?*

Mission Patches

The Principia mission patch was designed by 13-year-old Troy Wood, who explains: "Principia refers to Isaac Newton's principal laws of gravity and motion, so I drew an apple because that is how he discovered gravity. Plus, Tim is promoting healthy eating as part of his mission and apples are healthy." Fittingly, a stylized space station glints in the apple. The Soyuz rocket taking Tim into space flies over the UK as the colors of the Union Flag run along the border. The design was the winner of a BBC TV Blue Peter competition, which received over 3000 entries.

The mission name, Principia, was also the result of a competition launched in 2014, with over 4000 entries, 20 of which were Principia, referring to Sir Isaac Newton's text *Philosophiæ Naturalis Principia Mathematica* (Mathematical Principles of Natural Philosophy). Published in three parts in 1687, it states Newton's laws of motion, including the law of universal gravitation. Peake himself said "I am delighted with this name that honours one of Britain's most famous scientists. I hope it will also encourage people to observe the world as if for the first time – just as Isaac Newton did. Our planet Earth is a precious and beautiful place and we all need to safeguard it."

The Soyuz TMA-19M patch was designed by Luc van den Abeelen and follows a very simple design showing the Soyuz spacecraft orbiting a blue Earth with three stars, presumably representing the three crewmembers. As usual the names and national flags of the crew surround the image.

Fig. 7.2. Principia mission patch (ESA)

Fig. 7.3. Soyuz TMA-19M mission patch (ESA)

Fig. 7.4. Expedition 46 mission patch (ESA)

Fig. 7.5. Expedition 47 mission patch (ESA)

The original Expedition 46 patch was designed with input from commander Scott Kelly, but was not well received for a number of reasons. The stylized 46 was unbalanced but also resembled both a swastika and the logo of a Ukrainian far-right paramilitary body, the Azov Battalion. For that reason, the stylized 46 was replaced with a simpler image. Earth is depicted at the top with the flags of the countries of origin of the crew members, and the names of the six Expedition 46 astronauts and cosmonauts are shown in the border.

The design for the Expedition 47 patch was created by artists Tim Gagnon and Jorge Cartes, which is based on a stylized Sokol spacesuit helmet visor and shows the view of the International Space Station that would be seen by the crew as they approach in their Soyuz. The station image had been drawn by Erik van der Hoorn for the Expedition 41 design, but had not been used. Like the previous patch design an initial stylized 47 was replaced with a standard figure to avoid similarities to political symbols.

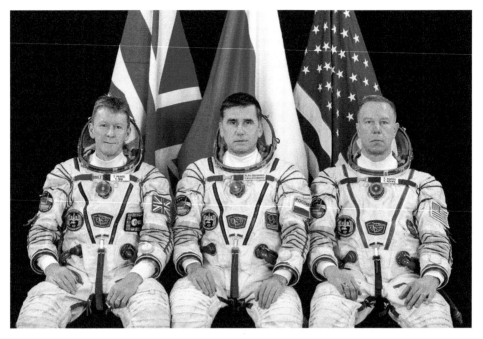

Fig. 7.6. The Soyuz TMA-19M crew with Tim Peake on left (www.spacefacts.de)

Fig. 7.7. The Expedition 46 crew with Tim Peake second from right (NASA)

Fig. 7.8. The Expedition 47 crew with Tim Peake third from right (NASA)

Timeline

Expedition 46, Week Ending December 20, 2015

Soyuz TMA-19M launched from Baikonur Cosmodrome, Kazakhstan at 11:03 UTC on December 15, carrying Tim Kopra of NASA, Tim Peake of ESA and Yuri Malenchenko of Roscosmos to the International Space Station. Soyuz Commander The automatic docking failed and Malenchenko manually docked the Soyuz TMA-19M spacecraft to the Rassvet module at 12:33 p.m. EST. The hatches were opened at 19:58 GMT and the Soyuz crew members were welcomed by Expedition 46 Commander Scott Kelly of NASA and Flight Engineers Mikhail Kornienko and Sergey Volkov of Roscosmos.

Like Andreas Mogensen and others before him, Peake started to fill out the series of questionnaires as part of the Space Headaches program. Twenty-four astronauts will ultimately participate in a study to evaluate headaches suffered especially in the first week in orbit. The form is filled every day for the first week, then every week thereafter.

After the routine safety briefing, the new crew members got to work. Peake worked on NanoRacks equipment and life support hardware and reviewed safety procedures. Together with Tim Kopra and Scott Kelly he started to unpack cargo from the Cygnus-4 spacecraft that had arrived on December 8, 2015. He was also introduced to the Advanced Resistive Exercise Device (ARED) and the station's exercise bicycle. In the first week, he started working on the Canadian Space Agency bone marrow study investigating the effects of radiation on the bone marrow and conduct his first weekly Crew Conference with ESA ground control at the Columbus Control Center.

On December 18 Peake reviewed the upcoming spacewalk plan, where Kopra and Kelly will move a stuck mobile transporter outside the station. Tim's role involved helping the spacewalkers don their spacesuits and monitor their progress for mission control.

On December 19, Progress 60 departed the station, undocking from the Pirs module at 7:35 a.m. UTC, freeing up the docking port for the arrival of Progress 62, which launched from Baikonur 8:44 a.m. UTC. At this point there were four spacecraft docked to the ISS: Soyuz TMA-18M at Poisk, Soyuz TMA-19M at Rassvet, Progress 61 at Zvezda and a Cygnus at Unity.

Expedition 46, Week Ending December 27, 2015

The Progress 62 cargo craft launched at 3:44 a.m. EST from the Baikonur Cosmodrome in Kazakhstan on December 21. On the same day, Kelly and Kopra started their spacewalk at 7:45 a.m. EST. The main purpose was to move the space station's mobile transporter rail car a few inches from its stalled position so it could be secured before the arrival of the Progress 62. They released brake handles on crew equipment carts on either side of the space station's mobile transporter rail car and latched them in place. With that complete, they performed so called get-ahead tasks, such as securing a second pair of cables in preparation for International Docking Adapter installment work to support U. S. commercial crew vehicles, routing an Ethernet cable for future connection to a Russian laboratory module and retrieving tools that had been in a toolbox on the outside of the station, so they can be used for future work. The EVA lasted 3 hours and 16 minutes and was the 191st in support of assembly and maintenance of the station.

Progress 62 arrived at the station on December 24, docking automatically with the Pirs docking compartment at 5:27 a.m. EST. It held 2.8 tons of food, fuel and supplies.

ESA Eurocom Andrea Boyd explains how Tim Peake experienced Christmas on the station in the ESA Principia blog:

> *Day 10 of the Principia Mission and it's been a much more intense time than new space station crew normally experience when arriving on the International Space Station! Tim has taken this in his stride and performed marvelously.*

> *Arriving on Tuesday 15 December, new crew are supposed to have some extra time during the first week to adapt to space as well as working, running the various science experiments and of course exercising. Tim's exercise includes continuing training for his upcoming London marathon. Their first day on*

board also had a one-hour emergency response practice drill, reviewing and performing in space things that they have practiced as a crew on ground many times. After that Yuri, Misha and Sergei went straight back to work in the Russian Science Laboratory while the two Tims and Scott returned to running many different science experiments in the Japanese Kibō, American Destiny and European Columbus Orbiting Laboratories.

Friday 18 December we had the first inflight call from space to the European Astronaut Centre – my colleagues and I enjoyed seeing Tim doing so well, answering questions from our visiting journalists and attempting his first live backflip! You can watch the full replay online.

On Tim's fourth day in space they realised an emergency spacewalk was needed, and crew worked all weekend to prepare this for Monday. Less than a week after launch Tim was tasked with being Intervehicular Support, which is an incredibly demanding role: to be responsible for the airlock, the status and the safety of both spacewalkers. On Sunday Tim was able to have some face to face time in his first weekly private family videoconference.

On Monday I was on console in the Columbus Control Centre to witness the Expedition 46 crew square away the critical task of this spacewalk – rescuing the stuck Mobile Transporter (MT). It was a race against time because a Progress cargo ship was launched on Monday, and it had to dock on Wednesday, so the Mobile Transporter needed to be fixed before then. The team aimed for Monday with Tuesday as a backup just in case. Thankfully, Scott and Tim Kopra solved issues with the Mobile Transporter just 20 minutes after leaving the airlock and then took on several tasks from spacewalks scheduled for next year. After finishing all of those, Tim guided them back inside the airlock, performed the ingress procedure and after all these checks were clear, opened the hatch to welcome them both back and get them out of the suits.

It's not quite like in the film Gravity *to doff the Extravehicular Mobility Units (EMUs): removing spacesuits requires extra assistance and takes about half an hour per suit. Sergei Volkov came to the airlock to assist for some of the procedure during his free time even though he wasn't scheduled to, he joined by choice just before the two spacewalkers returned inside to make Tim's tasks faster and to help the whole crew. The International Space Station crew is fantastic like that – we can clearly see the camaraderie from mission control.*

A few days earlier the Progress 60P cargo ship had undocked from the station, taking out rubbish with it that burnt up in the atmosphere as planned, freeing up the port for the next Progress 62P cargo ship, which launched the same Monday as the spacewalk and chased the space station for a couple of days, docking automatically to the same port on Wednesday. Christmas presents, new science, fuel and food supplies were welcomed by the crew, and Yuri, Misha and Sergei are unpacking about 3 tonnes from Progress while Tim, Tim and Scott are continuing to unpack about 3 tonnes from the Cygnus cargo ship.

A metric tonne each is a lot for an astronaut to unpack around their science and other duties, so it takes a while. Amongst the new experiments brought up is one of my favourites called BASS, which stands for Burning and Suppression of Solids – it's basically setting controlled fires to test and develop flame-retardant textiles. Fire in space is one of the three potential onboard emergency situations, but in a controlled experiment container like BASS it looks phenomenal!

It's been a pleasure to chat with Tim during various science experiments. You can listen to our entertaining British-Australian accent clash again next week as well as hear all the other astronauts chat with different Mission Controls: the space to ground channels are broadcast live all the time on www.ustream.tv/ channel/live-iss-stream. To distinguish between the two Tims we first call up using their surname, Kopra or Peake, and then switch to just saying Tim.

The Columbus Control Centre had various Christmas touches subtly added around the consoles, and the Mission Directors at the start of this month left each Flight Control Team console an advent calendar or three (depending on shifts) to open up a door of chocolate each day in December. I added a Koala bauble tree and Christmas turtles, fun space fact: Houston mission control and Munich Mission Control both have lakes with turtles in them, and Moscow mission control sent turtles orbiting around the Moon long before Apollo. We have a running joke that all successful spaceflight programs need turtles!

On Christmas Eve the UK Space Agency and Tim surprised us with presents delivered to the control room! The Christmas care package included mission goodies and completely amazing, brilliant Christmas food and items that I'd never expected to see this year! Fruit Mince Pies!! Cadbury Chocolates!! Christmas Crackers!! There's a few Brits in the team who appreciated these in particular as did I, having grown up in Australia, which is heavily influenced by British culture. Also some copies of the fabulous new Astronaut Handbook (must confess that I own the extended version of this already and it's one of my favourite books, written after consultation with Tim and the ISS Flight Control Team).

The crew has Christmas Day itself off as a well-deserved rest and a bonus one-hour video conference each to talk to their families. The Operations Planning team made some colourful modifications to our usual timeline on December 25. Saturday as usual the crew perform cleaning, chores and maintenance tasks around the space station (don't think that being an astronaut excuses you from chores! They divide up the kitchen, bathroom, vacuuming chores and so on just like you do at home!). Sunday is the astronauts' only day off each week, including a one-hour videoconference call to their family and Monday to Friday is back to 12-hour working days as usual.

Did I mention Tim sent us Fruit Mince Pies!!?? I was VERY excited, and my taste buds were dancing. Hoping there's still some left for seconds on my next shift…!

Expedition 46, Week Ending January 10, 2016

On January 4 Peake started work on NASA's Advanced Colloids Experiment-Heated-(2ACE-H2) experiment. This investigated self-assembling materials with colloids, small particles suspended in a liquid. He recorded the microscopic particles settling without the hindrance of gravity. Later, he initiated the Canadian Vascular Echo experiment which examines why astronauts' arteries age and become more rigid in space.

Peake and Kopra prepared, on January 5, for an upcoming spacewalk to replace a failed voltage regulator. Kopra worked on the U. S. spacesuits and Peake collected and configured their spacewalk tools. Peake said "I am thrilled at this opportunity for a spacewalk. Right now, we are focusing on preparing the tools, equipment and procedures. Maintaining the International Space Station from the outside requires intense operations – not just from the crew, but also from our ground support teams who are striving to make this spacewalk as safe and efficient as possible. If the spacewalk is successful, this will restore the International Space Station to 100% of its operational capability." The station has eight shunt units to regulate power but had been operating with only seven since November 2015. He also used the station's ham radio for the first time to talk to a school in the United States.

Peake and Kopra (the two Tims) checked out their U. S. spacesuits, ensuring a good fit, and readied their spacewalk tools. The crew joined Scott Kelly, as he observed a moment of silence on January 8 to mark the fifth anniversary of the shootings in Tucson, Arizona. Congresswoman Gabrielle Giffords, married to Kelly's twin brother, astronaut Mark Kelly, was gravely wounded in the attack. Scott Kelly was in space commanding Expedition 26 that day. Peake also spoke to a school in the UK over the ham radio and worked on preparing more experiments in the Japanese Kibō laboratory.

Expedition 46, Week Ending January 17, 2016

The new week got underway with Peake and Kopra continuing to prepare for their spacewalk. Peake also joined Kopra, and Malenchenko reviewed medical training and CPR procedures. Kelly and Peake set up the JAXA Electrostatic Levitation Furnace (ELF) equipment. It is installed into the Multi-purpose Small Payload Rack 2 (MSPR2) work volume inside of the Japanese Experiment Module (JEM). Tim wrote in his ESA blog on January 14:

> *Tomorrow Tim and I are going on our EVA. We have been preparing for this specific spacewalk for weeks in space, and months before that on Earth. However, to undertake an EVA actually takes several years of training. We have spent many hours working in our spacesuits, 'floating' in the largest swimming pool on Earth with a space station mock-up, have used virtual reality headsets to re-enact our operations and trained for the worst case scenario of becoming detached from the space station, but I guess nothing can fully prepare for the feeling of being outside of a spacecraft in the vacuum of space.*

Although I am exhilarated by tomorrow's spacewalk, I have no time to dwell on these emotions. The six hours and thirty minutes we will work on the space station's hull are meticulously planned, and Tim and I need to execute each step methodically.

In the past few weeks we have been preparing our tools and going over the EVA timeline that is almost 40 pages long. Reid Wiseman will be talking to us and guiding us from mission control – all operations in space rely on tremendous support from the dedicated international ground operations team. Reid has two spacewalks under his belt, and his second EVA was similar to ours in that he swapped a Sequential Shunt Unit (SSU) – we cannot think of a better colleague to talk to on the space-to-ground voice loop.

Our tools and spacesuits are ready, with all of our tools either clipped onto our spacesuit's 'Mini Work-Station' or stowed inside tool bags in the order we need them. In previous EVAs the bolt that keeps the SSU didn't always turn smoothly, and NASA thinks this might be because the thread gets blocked with debris. In true Martian style we fabricated a makeshift tool out of a toothbrush to clean the pinion thread if necessary.

You stop, you drop.

I can hear my trainers at the European Astronaut Centre and their constant drilling in my ears: "you stop, you drop" meaning that as soon as you stop moving from A to B you 'drop' a tether – a short strap securing you to the nearest handrail. In space, if it isn't fixed down it will float away, and that includes ourselves. As we move to the furthest edge of the space station, we will be attached to an anchor point by a thin steel wire on a reel, called a Safety Tether. These thin steel wires are a double-edged sword, however, as we must remain vigilant to not get them tangled up.

As soon as we exit the airlock, we will keep check on each other. The helmet in our EMU suits does not move, so I rely on Tim to check nothing is caught or snagging, as Tim relies on me to check his back. Spacewalks, like many critical operations, operate on the buddy-buddy system. Tim and I will constantly be checking each other and relying on each other for assistance if something should go wrong.

So having completed all of our training and preparations – it's finally time to go for a walk. See you on the other side of the airlock; we have already packed our toothbrush!

Tim Kopra and Tim Peake started their spacewalk when they switched their spacesuits to battery power at 7:48 a.m. EST on January 15. They planned to replace a failed voltage regulator. The Electrical Power System consists of an acre of solar panels that take in sunlight to generate, store and distribute power. After successfully replacing the regulator, after 2 hours, they continued with additional tasks, including the routing of cables in advance of International Docking Adapter installment work to support U. S. commercial crew vehicles. The EVA ended at 12:31 p.m. EST.

Fig. 7.9. Tim Peake on his first EVA, January 2016 (NASA)

Once again, an astronaut, this time Kopra, reported a water bubble had formed inside his helmet. Kelly helped the spacewalkers to remove their spacesuits and helmets. Then they used a syringe to take a water sample and retrieve the helmet absorption pad to determine what may have prompted the water to form inside Kopra's helmet. This was the 192nd EVA in support of assembly and maintenance of the orbiting laboratory.

Expedition 46, Week Ending January 24, 2016

After the terminated spacewalk, the crew examined the spacesuit. The leaking water was stored for late investigation. The suit repressurized and tested no leaks were found.

On January 20, Kopra and Peake continued the long-running Ocular Health vision study by taking eye and blood pressure readings. Kopra then scrubbed spacesuit cooling loops while Peake installed the equipment for the Airway Monitoring experiment to determine lungs are affected by microgravity. The following day, the crew conducted an emergency drill. The tested and validated communication protocols. They familiarized themselves with safety equipment and procedures and memorized escape routes. There was a conference call with ground controllers afterwards.

On January 22, Kelly and Kornienko had completed 300 consecutive days in space as part of the Year in Space mission, which had begun in March. Even before the mission was completed, this gave Kelly the record for the longest American spaceflight.[1]

Expedition 46, Week Ending January 31, 2016

On January 25, Peake worked on the Electrostatic Levitation Furnace, a device that levitates, melts and solidifies materials to study the thermophysical properties of different metals. He later tested touch-based technologies and repaired sensitive equipment. Volkov and Yuri Malenchenko resized their Russian Orlan spacesuits, checked them for leaks and set up hardware before their maintenance spacewalk.

On January 27, Kelly and Peake loaded a satellite carrier and its deployer mechanism in the Kibō laboratory's airlock in anticipation of the launch of the Aggiesat4 and BEVO-2 nanosatellites. These university nanosatellites are to refine autonomous navigation, rendezvous and docking software and procedures. Peake and Kopra then stowed rubbish onto the Orbital ATK Cygnus to eventually burn up on re-entry.

---

[1] Kelly would ultimately spend 340 days on this single mission, still a record for an American. Next longest American mission is Peggy Whitson's subsequent 289 days during Expeditions 50, 51 and 52. The global record is still held by Valeri Polyakov, who spent 427 days on board Mir in 1994/1995 as part of a Mars mission analog.

The following day, the crew observed a moment of silence to remember the sacrifice of the crews of *Apollo 1* and space shuttles Columbia and Challenger.

ESA flight director Tom Uhlig wrote about Tim Peake's work on the station in the ESA blog:

You don't need to have studied aerospace to work in the Flight Control Team. I am a physicist for example and did research in electron-microscopy, looking at the smallest particles at the atomic level. To operate the huge, room-sized million-euro machine needed for our research we undergraduates at the university in Regensburg had to obtain a "driving license" before we were allowed to take nice photos of silicon atoms. To operate the microscope at its limits is not easy: the electromagnetic lenses must be perfectly adjusted, the sample must be well-prepared, placed in a good vacuum with a stable temperature (liquid nitrogen ensured a stable cold) and most of all: any vibrations must be eliminated. Our microscope was mounted on a concrete foundation several tons in weight in the basement of our university – separate from the rest of the building. Even an enthusiastic student running up stairs somewhere in the building could ruin a 'photo shoot' and the research. Some labs such as our colleagues in Dresden do electron holography microscopy in the middle of forests so city-vibrations don't get in the way. Of course, absolute silence in a room is needed – an electron is a highly sensitive microphone, but in a negative way…

Who would have thought that the Space Station drifting weightlessly around Earth has similar problems? Space research specifically uses the term 'microgravity' for Space Station research, not 'weightlessness,' as there is gravity just several orders of magnitude less than on Earth. Gravity still acts at the altitude of the space station – but the station is flying at a speed that the centrifugal force compensates for it. Strictly speaking this compensation only applies at the exact centre of gravity so small uncompensated forces occur in the huge structure further away from the centrifugal epicentre.

Other effects occur as the station decelerates due to atmospheric drag. And then there are vibrations. The space station is a highly complex machine, inside pumps pump, fans rotate, humans work and robots move. All these mechanical moments lead to vibrations in a variety of frequencies – and their effect grows with increasing distance from the centre of mass of the station.

In the European Space Agency's Columbus space laboratory, therefore, we arrange the experiment cabinets by sensitivity to "micro-g environment disturbance." Everything that is very sensitive is located near the entrance hatch – closest to the space station's centre of mass. Here we examine liquids in microgravity and grow crystals. On the opposite side of Columbus, we do the less sensitive research: here we have the racks for human physiology – vibrations don't have a major influence on the metabolic processes in human bodies, for example.

In the future we will install a delicate experiment on the less favourable side of Columbus' exterior, the Atomic Clock Ensemble in Space (ACES). ACES will test highly accurate atomic clocks in space and refine Einstein's theories.

Before we fly the experiment, we want to get the most accurate values on microgravity conditions, therefore we added some preparatory activities to Tim Peake's schedule: First we had Tim Peake rearrange some items in Columbus, to access the end cone region. Then he took one of the Space Acceleration Measurements System (SAMS) sensors from the Japanese Kibō module and fixed it to the end of the Columbus laboratory. This sensor system allows us to measure accelerations, forces and vibrations and will provide us with valuable information in the next weeks. Our team followed Tim's work on the camera – thanks to the good preparation, the Brit could do his job without complications and questions.

The measurements will begin soon and pave the way for Columbus to tick to the sound of very precise clockwork. Time is of the essence …

Expedition 46, Week Ending February 7, 2016

On February 1, Peake retracted a small satellite deployer back in the Kibō laboratory module and performed some maintenance work on the BioLab incubator. Later, he collected and tested samples from water dispensers in the U. S. and Russian segments of the orbital laboratory. The samples would be returned to Earth inside a Soyuz spacecraft for further analysis. He conducted his third session of testing his own skin for the Skin-B experiment. He was the fifth astronaut to perform this task.

Malenchenko and Volkov began a planned 5.5 hours spacewalk from the Pirs Docking Compartment at 7:55 a.m. EST on February 3. They retrieved the EXPOSE-R Experiment, a collection of biological and biochemical samples placed in the harsh environment of space. To this point 46 species of small organisms and over 150 organic compounds have returned after spending 18 months bolted to the Zvezda module. This is part of ESA's research into astrobiology, or the study of the origin, evolution and distribution of life in the universe. They concluded their spacewalk at 12:40 p.m. EST. Lead investigator Hervé Cottin, from the University of Paris-Est Créteil, explained "Exposure is nothing less than a little chemistry box to help us better understand chemical reactions in space. If a molecule survives 18 months in space then it could come to Earth from space. If a molecule has changed after its 18 months voyage then we know that space travel filters our observations on Earth and a chemical might have formed from a different configuration."

Peake helped Malenchenko and Volkov as he stowed the U. S. equipment used on the suits. Peake later videotaped himself reading a children's storybook and performing science demonstrations for students. He also switched on the small computer 'Astro Pi' which executed student-developed code. He had two such computers, named Ed and Izzy.

Station commander Kelly and Peake worked in the Kibō laboratory stowing a Cyclops satellite deployer and checking for leaks in the airlock. On Friday he participated in the NASA Fine motor experiment. In the afternoon he operated the combustion experiment facility to prepare it for more experiments. On Saturday, he spoke to ham radio operators and saved a stool sample for the Japanese Multi-omics study. This examines how spaceflight affects 'good' bacteria in the body. The sample was frozen in the MELFI freezer for return to Earth. After this less than glamorous activity in the name of science he and his colleagues had a treat as they watched the England vs. Scotland Six-Nations rugby match.

Expedition 46, Week Ending February 14, 2016

Peake wrapped up maintenance work on the Electrostatic Levitation Furnace on February 8. He also explored crew immunology to potentially improve astronaut health and life on Earth. He then joined Kelly in the Quest airlock to get a U. S. spacesuit ready for gear replacement work. He spent time the Kibō laboratory working on Multi-Omics.

On February 9, Kelly, Peake and Kornienko each participated in a different experiment looking at the immune system of space residents. Kelly collected body samples looking for microbes that could potentially cause infections or allergies and stowed them in a science freezer for analysis. Peake took a saliva sample for an experiment that is researching biomarkers for immune dysfunction in space. Kornienko explored how radiation and other unique factors of living in space could affect a crew member's immune system. Kopra and Peake also partnered up for ultrasound scans of their arteries with guidance from doctors on the ground. The ongoing Cardio Ox study looks at an astronaut's carotid and brachial arteries before, during and after a space mission.

Peake organized a competition based on his music playlist. Each astronaut can bring up 14 GB of music to space. The #SpaceRocks competition involves Peake Tweeting the line of a song, and the public can reply with the song title and artist. He organized over 60 patches as prizes. The responses were so numerous and fast that the winners could not be determined. The rules were changed to randomly select one winner from the correct answers received in the first 20 seconds.

Fig. 7.10. SpaceRocks patch (ESA)

Peake attached sensors to himself on February 10, so doctors can monitor how a crew member adapts to 16 sunrises and sunsets a day while in low-Earth orbit. He also spoke to French television France 2.

Kelly and Peake completed the replacement of parts on a U. S. spacesuit on February 11. Ground controllers moved the Canadarm2 into position for the upcoming grapple and release of the Cygnus cargo ship attached to Unity, before it is deorbited.

On Friday, he completed a session of recording sounds. He wore a microphone for 24 hours and had placed two microphones around the Station to record ambient noise. In the evening Tim did an amateur radio contact with the Royal Masonic school for girls in Britain that included video over simple radio waves.

Expedition 46, Week Ending February 21, 2016

On February 15, Peake drew blood and took eye readings as part of the Ocular Health experiment. He later worked on the Electrostatic Levitation Furnace in the Kibō laboratory and tidying in the ESA Columbus laboratory.

The following day, he had more exams. He installed an Astro Pi in the Node-2 module as part of the educational program. He also recorded a video on the CardioVascular health experiment for the Canadian Space Agency. He worked on the water supply and carried out maintenance on the Microgravity Acceleration Measurement System. This records vibrations and the effect they have on experimental equipment.

On February 18, Peake and Kopra examined how astronauts work on detailed interactive tasks and measured cognitive performance. Peake also studied the thermophysical properties of different metals inside the JAXA Electrostatic Levitation Furnace.

Kelly and Kopra unberthed the Cygnus cargo ship with the Canadarm2 robotic arm at 7:26 a.m. EST over Bolivia. Earlier, ground controllers at NASA's Johnson Space Center had maneuvered Cygnus into place for its departure. At the same time, Peake worked on the Fine Motor experiment, had his weekly conference with ESA ground control and exercised.

On February 19, The UK space agency, ESA, and the Raspberry Pi Foundation launched a competition for UK schools. The students were asked to write code for Peake's two Astro Pi computers. A previous competition saw seven winning programs launched as part of the Astro Pi payload. Peake deployed the first Astro Pi in the Columbus laboratory on February 2 and the second on February 16 in the Harmony Node 2 module running Earth observation experiments, looking through the nadir hatch window. Peake gave the students two challenges: write Python code to turn the Astro Pi into an MP3 music player and compose music using a tool called Sonic Pi. Peake said "This competition offers a unique chance for young people to learn core computing skills that will be extremely useful in their future. It's going to be a lot of fun!" Libby Jackson from UK space agency said: "We are excited that the Astro Pi project is being extended to allow more students the opportunity to see their code in space. There were some fantastic ideas in the first competition and I am sure that the new challenges will see more interesting ideas being submitted." David Honess from Raspberry Pi Foundation said: "Tim told us that the software for updating his MP3 player is not approved for the ISS laptops, so he's potentially facing another 4 months without any new music. So there's a practical, utilitarian purpose for having the students code this MP3 player for him. It'll solve a real problem on the space station."

He carried out a radiogram and replaced hard drives on the Multi-Purpose Small Payload Rack computers. He did a session on the Japanese Microbe-IV study and worked on the space station's spacesuits. After lunch he worked on the Kibō's power system. Cygnus's departure resulted in a need for rerouting some cables in the station.

Expedition 46, Week Ending February 28, 2016

On February 22, Peake worked on the Japanese Airlock and then worked on the Advanced Colloids Experiment looking at how microscopic particles behave in liquids.

On February 23, Peake replaced cables in the station's Advanced Resistive Exercise Device and installed the NASA Space Automated Bioproduct Laboratory (SABL), which is capable of supporting life science research on microorganisms, small organisms, animal cells, tissue cultures and small plants. Later, he checked out new science equipment that will support new biology research and set up hardware for an experiment that explores the risk of airway inflammation.

On February 24, Peake wrote in his ESA blog:

Breathe in, breathe out. Tim Kopra and I are heading back to the airlock... This time we are not going on a spacewalk but instead we will be pumping some of the air out of the Quest module in the name of science. By reducing the pressure in the airlock, we can simulate being at high altitude – or in a future lunar habitat that will likely be kept at lower pressures than on Earth – because it is cheaper and easier to build structures to withstand lower pressures.

Living in space is not easy on the human body. Over millions of years humans have evolved to live with gravity; take it away and our bodies adapt and cope remarkably well, but I still need to monitor almost every aspect of my health up here. Luckily I have a flight surgeon at the European Astronaut Centre and a team of biomedical engineers who watch out for me daily, but space medicine is still a very new discipline.

One area researchers are looking at is my lungs – astronaut's lungs may become easily irritated or inflamed if we inhale dust particles (thankfully I haven't had any problems so far). For the Airway Monitoring experiment Tim and I will breathe into a specially developed mask that measures the nitric oxide we exhale which is a good indicator for inflammation of the lungs.

The weightlessness makes dust more likely to get into our lungs. On Earth, dust settles on the floor, but without gravity dust never settles, which is one of the reasons we clean the space station as thoroughly as we can every Saturday. On the Moon and Mars, it will be worse because, although there is gravity, it is weaker than on Earth, and there is a lot more very fine dust, which also sticks to astronauts due to static electricity.

On Earth doctors would use an X-ray or CT scan to test whether a patient's lungs are inflamed, but in space these large machines are not an option – so necessity breeds creativity. The researchers of the Airway Monitoring experiment devised this "simple" nitric-oxide test and Tim and I are part of the study to see if it works in space.

The Airway Monitoring experiment is the kind of research we astronauts love. Not only do we get to use the airlock for scientific purposes – but we are also contributing to creating knowledge that will help our future colleagues explore new environments in our Solar System. Moreover, this study will increase our general understanding of problems with airway inflammation and could provide valuable research into new methods for treating conditions such as asthma. As a bonus Airway Monitoring does not require that we draw blood samples, so the needles can stay in their pouches!

Breathe in, breathe out. For most people this is second nature and we hardly think about it until something goes wrong. For the millions of people suffering from asthma around the world this nitric-oxide lung test could offer a quick and cheap way to diagnose lung problems everywhere – if it works in space it should work anywhere on our planet.

Kopra and Peake successfully undertook the Airway Monitoring experiment on February 25.

Expedition 46/47, Week Ending March 6, 2016

On February 29, Tim Kopra took over command of the ISS from Scott Kelly in a traditional Change of Command ceremony. Meanwhile, Peake worked inside Japan's Kibō laboratory module preparing to increase its stowage capacity.

Expedition 47 began when Scott Kelly, Mikhail Kornienko and Sergey Volkov undocked from the station at 8:02 p.m. EST to begin their voyage home with Volkov in command and at the controls. This marks the end of the One Year mission for Kelly and Kornienko who each logged 340 days in space across Expeditions 43, 44, 45 and 46. They landed in Kazakhstan at 11:26 p.m. EST.

Back on the station, Peake worked in the Columbus module checking out science hardware for a magnetic field experiment and a payload transfer rack. He had another session with the Fine Motor Skills experiment and worked on the sensor unit for the Electrostatic Levitation Furnace. He made a video recording of his daily activities while checking his microbiome as part of the JAXA Multi-Omics study. In the afternoon he removed JAXA radiation detectors to be returned to Earth for inspection. He joined Kopra and Malenchenko in reviewing emergency procedures.

After the departure of the Soyuz, Peake helped organize blood samples for NASA's twin study with Scott Kelly and his twin brother Mark. He photographed the station interior and worked on the MagVector experiment. He maintained the Urine Processing Assembly and Water Recovery System.

Expedition 47, Week Ending March 13, 2016

Tim Peake started installing and routing cables that will enable communications with future commercial crew vehicles on March 7. The Common Communications for Visiting Vehicles, or C2V2, consists of both radio frequency and hardline connections that will be used during rendezvous, docking and mated activities at the space station.

The following day, he set up the Microgravity Science Glovebox for Rodent Research operations. That experiment was due to start after the arrival of the next SpaceX mission.

The crew practiced the communication and procedures necessary to escape an emergency situation. They practiced departing the space station quickly and entered their docked Soyuz spacecraft for use as a lifeboat. Before the drill, Kopra and Peake participated in a series of tests on a touchscreen tablet for the Fine Motor Skills study. The experiment helps researchers understand how astronauts concentrate and work on detailed tasks and sensitive equipment during and after a long-term space mission. Peake then moved on to the Magvector experiment and studied magnetic fields and electrical conductivity, possibly setting up the space station for future astrophysics research. He also had an interview with Phillip Schofield and Holly Willoughby for ITV's This Morning show.

Kopra and Peake continued to install hardware that would communicate with future commercial crew vehicles on March 10. The equipment will enable hard-line and frequency communications with the private spacecraft during rendezvous, docking and mated activities. Peake spent a few moments collecting a saliva sample for a study that observes the human immune system in space. He is also helping engineers understand the factors necessary for a comfortable living space during long term missions. Kopra spoke over ham radio to school in North Dakota in the United States. This was the 1000th organized Amateur Radio on the International Space Station (ARISS) event.

Expedition 47, Week Ending March 20, 2016

Peake started the week running the Magvector electromagnetic study for the last time. In the Kibō laboratory, he replaced Payload Data Handling hardware required to run upcoming life science experiments.

On March 15, he joined Kopra for blood pressure checks. They used a tonometer and were monitored by doctors on the ground. Peake also tested and examined the COLBERT treadmill located in the Tranquility module.

Peake wrote in his ESA blog:

I have always been fascinated by remote-controlled vehicles. Like many kids I looked in amazement at remote-controlled cars and boats in the park and was eager to have a go whenever I could. As a young teenager I loved building, flight-testing and all–too-often crashing my radio-controlled aircraft. I learned the hard way that an aircraft's centre of gravity was rather important when it comes to stability and control! As an adult my passion for remote-controlled vehicles has not waned, which is why I'm particularly excited about an upcoming experiment where I get to drive a remotely-controlled rover on Earth…from here in space.

Remotely operated rovers and robots can be life-savers in disaster areas where humans cannot safely explore, such as searching for earthquake survivors or investigating nuclear disasters. Another area humans cannot easily explore is the surface of our neighbouring planets and moons, but controlling a rover millions of kilometres away poses many challenges. For example, a command sent to the Curiosity rover on Mars can take up to 24 minutes to reach the plucky rover, and the operators need to wait the same amount of time again to see the result of their command! This time delay makes for slow progress and does not allow real-time response to a rapidly changing situation.

ESA's Meteron project is looking at a way to not only reduce this time delay but also make the command path robust against signal interruptions that may occur over long distances with multiple relay stations. Astronauts en route or orbiting Mars could control a rover with much less time delay in order to scout new areas or prepare a space base in advance of humans arriving on the planet's surface.

On March 17 and April 25, I will be part of an experiment to drive two different rovers from Europe's Columbus laboratory. At first glance it might seem easy to operate a rover from space – the hard part would be getting astronauts and the rovers to their destination – but the communications network and robot interface need to be built from scratch and tested to work in microgravity.

On Thursday I will be driving Eurobot at ESA's technical centre in The Netherlands in a two-hour experiment. The space station flies at around 17,150 mph (28,800 km/h) and completes a full orbit of Earth every 90 minutes, so throughout the session my commands could be beamed directly to the car-sized rover while at other times the Space Station will be on the other side of the planet – so commands need to be relayed through satellites and multiple control centres. To keep everything going smoothly engineers at ESA's operations centre created a new type of 'space-internet' that adapts to the changing connections and network speed.

In the experiment this week Eurobot will be near a mock-up lunar base that has a problem with its solar array. It is up to me to locate, grab and unjam the solar panels.

Few astronauts have got their real-time 'rover wings,' so I am excited to be joining the exclusive club that includes NASA astronaut Sunita Williams, and my ESA astronaut colleagues Alexander Gerst and Andreas Mogensen who have already gotten their space-rover driving-license. This week's experiment is similar to Andreas's Eurobot experience during his Iriss mission last year but the Meteron engineers have added an extra challenge this time: they want me to fix the solar panel manually. Whereas Andreas commanded the rover to locate the solar panel I will be testing a new part of the software by finding the solar array interface and directing the rover to make contact.

In April the second Supvis experiment promises to be even more interesting as we will be upping the ante again. In coordination with the operations team at the ESA's mission operations centre in Darmstadt, Germany I will be driving a rover called Bridget at the Airbus Defence and Space "Mars yard" in Stevenage. The goal in this experiment is to navigate a landscape similar to a Martian cave or a Moon crater, but I will have had no foresight whatsoever for this task: the goal is to see how well an astronaut-robot system can adapt to unforeseen scenarios. For this experiment I am being kept in the dark – literally. To test the system to its limits the experiment will be run in the dark, as if it were exploring a real cave. Considering that typical rovers find it difficult to make sense of their environment in these conditions, this experiment is about putting the human in the driving seat for these challenging operations in the dark. Let's hope that my skills have improved somewhat since those early days of flying radio-controlled aircraft!

Kopra and Peake took part in the Ocular Health study on March 17. Peake worked on the technology that will control rovers on another planet from a spacecraft.

He wrote in his ESA blog on March 18:

In the movie 'The Martian,' Mark Watney was left stranded on Mars and survived for months by growing potatoes in Martian soil mixed with human waste as fertiliser. Thankfully, on the International Space Station the situation is not quite so dire, but food is critical for our survival nonetheless! Normally, at least six months' worth of food for a crew of six is maintained onboard the space station. Of course, none of this is fresh, nor can it be refrigerated or frozen. The food that we eat is either irradiated, rehydrated, canned or dried goods. Our 'kitchen' consists of an electrical food heater and a hot water dispenser…and that's pretty much it.

Feeding astronauts is not as straightforward as you might think. Of course, we require sufficient calories, vitamins, minerals and other nutrients to enable the human body to function normally. However, the human body undergoes a few changes in microgravity and we have to consider this in our diet. For example, too much salt can exacerbate the loss of bone density that astronauts suffer from in space, and much of our food can taste a bit bland as a result. Normally, astronauts take a vitamin D supplement each day to help counter the effect of being isolated from the sun's rays for six months. Additionally, 'space food' must be lightweight and have a long shelf life. Our standard Space Station menu takes all of these factors into consideration and does a pretty good job…but it is a bit repetitive, and you can often tell 'tastiness' was not highest on the list of priorities!

Thankfully, this is not the whole story – we get 'bonus' food! Although this sounds like we've won some sort of competition – bonus food actually accounts for around 10% of our calculated calorific intake and so it is vital to our well being. Astronauts are allowed to choose their own bonus food as long as it does not cost more than the standard food it replaces. When I was asked what bonus food I would like to take with me, I could have picked existing, commercial products, but it seemed like a much better idea to offer this up as a challenge to kids to get them thinking about healthy eating, nutrition and the problems associated with getting food into space. Oh – and I added one other very important criteria: tastiness!

The Great British Space Dinner competition was a huge success and engaged 2,000 students from both primary and secondary schools from all over the UK. Of course, every competition requires a prize, so it was decided that the winners would get to develop their menu with a celebrity chef to be used as inspiration for the actual meals that would be sent up into space as my bonus food. So, 10% of my diet was now in the hands of schoolkids...and they did not disappoint. The students had put so much thought and effort into their menus that it was hard to pick just three winners as was originally intended – so instead we ended up with five winning teams! Each menu had been carefully thought out as something fun, healthy, nutritious, tasty and with a hint of Britishness to give morale a boost and to help keep me connected with home whilst in space.

Having been on the International Space Station for over three months now, I can testify to the increased importance of the role that food plays in space. Our sense of smell is diminished in space because all the fluid shifts up to our chest and head, giving us a 'puffy face' look and often a feeling of nasal congestion. Our circadian rhythm can struggle with the sixteen day/night cycles we experience every 24 hours and this can also affect our appetite. We exercise for two hours every day and so require a good amount of calories, but often the food portions are small and so we have to be careful to monitor that we are eating a sufficient quantity and getting a balanced diet. Most importantly, living in a confined, artificial environment with a recycled atmosphere we feel the isolation from planet Earth – suddenly that bonus food takes on a whole new role in terms of morale and psychological well being.

So, to all the students who took part in the Great British Space Dinner challenge I would like to say a very special thank you from here in space for taking part and for developing such important fuel for my mission. Remember that healthy eating, exercise and a well-balanced diet is just as important for life on Earth as it is in space.

The Soyuz TMA-20M launched from the Baikonur in Kazakhstan at 5:26 p.m. EDT on March 18. Jeff Williams of NASA and Roscosmos cosmonauts Oleg Skripochka and Alexey Ovchinin were on board. A day later, at 11:09 EDT, the Soyuz docked to the Poisk module approximately 250 miles above the southern Pacific Ocean off the western coast of Peru.

Expedition 47, Week Ending March 27, 2016

Peake continued to take samples for the Japanese Multi-Omics experiment to look at how astronauts' microbiomes change in space. He prepared the combustion experiment rack to prepare for the Flame Extinguishment-2 experiment. In the afternoon he set up radiation dosimeters as part of the Habitability experiment. Later he joined Kopra as they had a training session simulating the berthing of the Cygnus spacecraft.

The ULA Atlas V rocket lifted off at 11:05 p.m. EDT on March 22 carrying the latest Cygnus cargo ship. On the station Kopra and Peake continued practicing using the Canadarm2 robotic arm to grapple and capture the spacecraft.

Peake spent March 23 running the SPHERES docking experiment. They were used to test and demonstrate advanced docking techniques, simulating spacecraft flying in space.

On Friday he transferred files for Astro Pi and had a conference call to prepare for the arrival of the Cygnus. He also filled in his weekly Space Headaches questionnaire and did a Habitability session.

Kopra, with help from Peake, successfully captured Orbital ATK's Cygnus cargo vehicle at 6:51 a.m. EDT on March 26. They, with help from the robotics officer in mission control in Houston, positioned Cygnus for installation to Earth-facing port of the Unity module. The cargo craft was berthed at 10:52 a.m. EDT, delivering more than 7700 pounds of science and research, crew supplies and vehicle hardware.

See Appendix 5 in this book: Tim Peake FAQ from ESA blog March 27, 2016.

Expedition 47, Week Ending April 3, 2016

Because Kopra and Peake worked over the weekend on the Cygnus grappling, they had a day off on March 28.

On March 30, the Progress 61 cargo spacecraft undocked from the aft port of the Zvezda Service Module at 10:15 a.m. EDT. Peake assisted Williams for ultrasound scans of his arteries for the Cardio Ox experiment.

Yet another cargo craft launched to the station on March 31. Progress 63 cargo craft launched at 12:23 p.m. EDT from Baikonur. At the time of launch, the International Space Station was flying about 251 miles over northeast Iraq. Less than 9 minutes after launch, the resupply ship reached preliminary orbit and deployed its solar arrays and navigational antennas as planned.

Peake swapped hard drives in a laptop computer that is recording data collected for a dark matter detection experiment. Progress 63 docked successfully to the rear port of the Zvezda Service Module at 1:58 p.m. EDT on April 2. It carried more than three tons of food, fuel and supplies.

Expedition 47, Week Ending April 10, 2016

The crew opened the hatches to Progress 63 on April 4 and began unloading nearly three tons of food, fuel and supplies. Peake carried out muscle research using specialised exercise equipment, Muscle Atrophy Research and Exercise System (MARES) and attached electrodes to his right leg and ankle. The crew conducted an emergency drill on April 6 in coordination with flight controllers. Peake practiced with the Canadarm2 in preparation for the arrival, capture and berthing of the next Dragon capsule. The SpaceX Falcon 9 rocket lifted off at 4:43 p.m. EDT on April 8.

On April 10 Peake, with the assistance of NASA's Jeff Williams, successfully captured the SpaceX Dragon spacecraft with the station's robotic Canadarm2 at 7:23 a.m. EDT. The ISS was traveling over the Pacific Ocean west of Hawaii. Dragon's arrival marked the first time two commercial cargo vehicles had been docked simultaneously at the space station. Orbital ATK's Cygnus spacecraft arrived just over 2 weeks previously. With the arrival of Dragon, the space station tied the record for most vehicles on station at one time, six. The Dragon delivered about 7000 pounds of science and research investigations, including the Bigelow Expandable Activity Module (BEAM).

Expedition 47, Week Ending April 17, 2016

Fig. 7.11. Tim Peake reading Yuri Gagarin's autobiography *Road to the Stars* on the anniversary of the first spaceflight, April 12. The book was signed by Gagarin, flew in space to Mir with Helen Sharman in 1991 and was signed by that Mir crew and the ISS Expedition 47 crew. (ESA)

Kopra and Peake continued vision tests and blood pressure checks for the Ocular Health study. The station was boosted again on April 11 with a series of burns, to prepare for the return home of Kopra, Peake and Malenchenko and the arrival of their replacements, Anatoly Ivanishin, Kate Rubins and Takuya Onishi.

On April 14 Peake worked on the Fine Motor Skills experiment, using a tablet computer to test crew performing interactive tasks. The crew carried out various biological studies, including looking at skeletal muscle cells under a microscope to examine atrophy, collecting saliva samples to check the immune system and testing cognitive performance during stressful conditions.

Peake talked to teachers in Norway, UK, and Poland. Teachers and students of all grades, as well as space scientists and engineers, gathered at national events taking place at York, Warsaw, and Oslo. The events were organized by ESA's ESERO UK, ESERO Poland, and Nordic ESERO offices. For the second part of the call Tim was joined by space station crewmates NASA astronauts Jeff Williams and Tim Kopra for an interview with Associated Press.

On April 16 the Bigelow Expandable Activity Module (BEAM) that arrived on the latest Dragon was installed at the aft port of Tranquility at 5:36 a.m. EDT. At the time of installation, the space station was flying over the southern Pacific Ocean. NASA and Bigelow are working together to test expandable habitats. "Expandable habitats are one such concept under consideration. They require less payload volume on the rocket than traditional rigid structures, and expand after being deployed in space to provide additional room for astronauts to live and work inside".

Fig. 7.12. Bigelow Expandable Activity Module installed on the Tranquility module (NASA)

Expedition 47, Week Ending April 24, 2016

The crew started the Genes in Space study on April 19. It is a student-designed experiment to study the linkage between DNA alterations and weakened immune systems in microgravity. The station's inventory was being updated after cargo transfers cargo to and from Progress 63 and Dragon. A U. S. spacesuit was prepared for return to Earth on the Dragon.

Later, the crew collected samples for the Plant Gravity Sensing-3 experiment. On April 24, Tim Peake completed a full 26 mile/42 km marathon on the station's threadmill in Tranquility. He started at the same time as the London Marathon.[2] He opened the London Marathon via a video message. His unofficial times:

Kilometers	Miles	Total Time	Split
5	3.10686	00:25:50	
10	6.21371	00:50:25	00:24:35
15	9.32057	01:16:36	00:26:11
20	12.4274	01:42:00	00:25:24
25	15.5343	02:09:30	00:27:30
30	18.6411	Routine loss of signal	
35	21.748	02:58:30	00:49:00
40	24.8548	03:27:08	00:28:38
42.195	26.2	03:35:21	00:08:13

Estimated start time: 09:00:00
Estimated stop time: 12:35:21

Peake wrote in the ESA blog on 24 April 24:

The run went better than expected. I thought I'd stick to a steady 7.5 mph, but when I got to 10 miles I realised that my legs were feeling OK but my shoulders were beginning to hurt, so I needed to finish the run quicker than planned and running faster doesn't seem to hurt the shoulders any more – in fact I think the longer stride made it less painful on the shoulders.

So I went to 8 mph for 10 miles and then for the last 6.2 miles went to 8.6 mph. My legs paid the price but my shoulders were grateful. It probably looked like I was having a strong run at the end but the reality was that I couldn't wait to get out of that harness!!

Watching the live marathon on the BBC the whole time was a huge encouragement – I had thought I would watch a movie (2001 A Space Odyssey was ready to go) or listen to my #SpaceRocks playlist, but in fact it was extremely

---

[2] He became the second astronaut to run a marathon in space, after Sunita William's Boston marathon run on the ISS in 2007 in Zvezda.

motivating watching the live coverage of the event and hearing the stories of some of the 33,000 people taking part. In addition to that I was able to compare my progress to the live event since I had the RunSocial app giving me an excellent view of streets of London as I would see them if I were running the real marathon.

Staying hydrated was no problem – this was a big mistake I had made in 1999 when I last ran the London marathon and did not drink enough during the race… which hit me hard at 18 miles and scuppered my plans for a sub three-hour run. This time I was drinking water from the start and I had my pouches lined up on Velcro strips on the panel above my head to ensure that I drank one pouch (300ml) per hour. I also had an energy sachet at 18 miles which was a great boost for the last stage of the run.

It was an incredible experience to take part in such a prestigious event whilst orbiting the planet on the International Space Station and I'm hugely grateful to everyone at the European Space Agency and NASA who made that happen. And last but not least, I was truly proud to be part of Team Astronaut in support of The Prince's Trust and to help raise awareness for the great work that they do.

Expedition 47, Week Ending May 1, 2016

Peake deployed a cubesat from the Philippines from the Kibō laboratory on April 27 for climate research. DIWATA-1, Filipino for "fairy" was a 50-kg-class microsatellite to observe Earth's climate to improve weather forecasting and natural disaster response.

Peake trained on the rover control software for the METERON experiment on April 28. Williams and Ovchinin continued the Fluid Shifts study. Kopra stowed equipment inside the SpaceX Dragon for retrieval back on Earth.

In order to maintain the station crew at six, it was announced on April 29 that the mission of Tim Peake, Yuri Malenchenko and Tim Kopra would be extended by 2 weeks to June 18. Peake said: "Although I am looking forward to being back on Earth and seeing friends and family again, each day spent living in space is a huge privilege and there is much work to do on the station. This extension will keep the Station at a full crew of six for several days longer, enabling us to accomplish more scientific research. And, of course, I get to enjoy the beautiful view of planet Earth for a little while longer!"

Peake collected tap water and other personal body samples for the Energy experiment on April 29. This examines the astronauts' energy requirements to improve health and performance. He also controlled a rover on Earth as part of the METERON experiment to learn how astronauts might control rovers on missions to Mars.

Expedition 47, Week Ending May 8, 2016

Peake swapped equipment on a specialized microscope that can download imagery and video to scientists on the ground. He also saved data collected from an armband for the Energy study, then moved on to the Rodent Research study that observes muscle and bone loss in space.

Robotics controller used the Canadarm2's cameras to survey the Alpha Magnetic Spectrometer on May 3. The following day, they started moving the robot arm to the Harmony module to be ready to grapple Dragon ahead of its release.

Expedition 47, Week Ending May 15, 2016

The SpaceX Dragon was loaded with scientific samples and physical results of experiments on May 9 and 10 before its release. The crew reviewed departure procedures and training for its release. The Russian cosmonauts had a day off on May 9 in observance of Victory Day, when Germany surrendered to the Soviet Union on May 9, 1945.

On May 10, the cosmonauts of the crew researched how space radiation affects human tissue using the human-like dummies as part of the Matryeshka study.[3] They also investigated how the carotid artery and immune system are affected by space.

The Dragon capsule was detached from the Harmony module and was released at 9:19 a.m. EDT on May 11. It then splashed down in the Pacific Ocean at 2:51 p.m. EDT, about 261 miles southwest of Long Beach, California. Williams, Kopra and Peake had a day of light duty after the departure of the Dragon, while the three cosmonauts continued their ongoing Russian research work and maintenance activities after taking Monday off to celebrate of Victory Day.

The Kibō laboratory's airlock was depressurized in preparation for the deployment of a series of nanosatellites.

Expedition 47, Week Ending May 22, 2016

On May 16, the International Space Station completed its 100,000th orbit. The first component, the Zarya cargo module, launched Nov. 20, 1998. Around the same time, cubesats were deployed from the Kibō laboratory's airlock as part of series of launches over 3 days. A laptop computer was readied ahead of the expansion of the Bigelow Expandable Activity Module (BEAM). The computer was to monitor sensors and prepare for upcoming BEAM operations.

---

[3] Matryeshka dolls are the traditional Russian nested dolls.

The cubesats deployed on the second day were Dove satellites built and operated by Planet Labs Inc. to take images of Earth for several humanitarian and environmental applications. A Waste and Hygiene Compartment (WHC) required maintenance and a new Advanced Recycle Filter Tank Assembly was installed.

By May 18, seventeen cubesats had been released from the small satellite deployer outside the Kibō experiment module's airlock. The latest groups of cubesats deployed provide Earth observations, improve commercial ship tracking and provide weather data on Earth's seas.

Williams scanned his leg with an ultrasound on May 19 for the long-running Sprint study. Cargo transfers continued to and from the Cygnus and the Canadarm2 robotic arm and DEXTRE scanned the RapidScat system that monitors weather patterns on Earth's oceans.

Expedition 47, Week Ending May 29, 2016

The crew started the week clearing up and packing the Rodent Research-3 (RR-3) experiment, which was completed the previous week. During the wrap up work, they also collected station air and astronaut breath samples for the CSA marrow bone study.

In preparation for a planned expansion of the BEAM module, the crew prepared the vestibule between BEAM and the rest of the station by pressurizing the area and performing leak checks. On May 25, Williams performed leak checks and installed hardware to monitor and support BEAM expansion. The first attempt to expand the BEAM module on May 26 were unsuccessful. Flight controllers informed Williams that BEAM had only expanded a few inches in both length and diameter.

Peake spoke with CNN International's Christiane Amanpour on May 24.

On the second attempt, on May 28, Williams pumped air into the BEAM module in a 22 seconds burst followed by an 8 seconds burst, to allow the module to stabilise and expand in between bursts. This was successful in expanding the module. He ultimately introduced 44 seconds of air during eight bursts. The module had expanded to its full size at 4:10 p.m. EDT. While over the south Pacific.

BEAM expanded from 7 ft long and just 7.75 ft in diameter packed to more than 13 ft long and 10.5 ft in diameter in its fully expanded mode. The capacity is now 565 ft^3 of habitable volume.

Expedition 47, Week Ending June 5, 2016

BEAM leak checks were conducted before hatches were opened. Further cubesats were launched from the Kibō module, supporting research such as communications and Earth observations sponsored by government, education and private organizations. The last batch was deployed on June 1.

On June 2, the BEAM pressure was being equalised with that of the rest of the station. Williams installed components on the BEAM bulkhead and vestibule area.

Expedition 47, Week Ending June 12, 2016

Williams opened the hatch to the Bigelow Expandable Activity Module (BEAM) at 4:47 a.m. EDT on June 6 and entered with Skripochka. They collected air samples and gathered data from sensors on the dynamics of BEAM's expansion. Williams said that the module was "pristine". Although the air was cold, there was no condensation on its inner surfaces. After the installation, the hatch was closed again.

Peake explored how astronauts adapt to tasks requiring high concentration and detailed procedures. Williams continued the Multi-Omics experiment looking at the immune system. Peake and Malenchenko loaded the Soyuz TMA-19M spacecraft and readied it to return to Earth.

The crew opened the hatch to BEAM again on June 7 as outfitting continued. The hatch was closed on June 8, with plans to keep it sealed until August.

Peake joined Williams and Kopra for ultrasound scans on June 9. Malenchenko, Skripochka and Ovchinin worked on their set of science experiments and maintenance tasks on the Russian side of the station. They continued exploring the vibrations the station experiences during spacecraft dockings, spacewalks and crew exercise sessions. They also researched new techniques to locate module pressure leaks as well as locate and photograph landmarks on Earth.

On the following day, Peake researched the cause of accelerated skin aging in space and studied plant hormones. Kopra drew a blood sample for stowage in a science freezer and later analysis. Skripochka researched the radiation the station and its crew are exposed to internally and externally. Ovchinin explored plasma physics while Malenchenko and Skripochka partnered up for cardiovascular health studies.

Expedition 47, Week Ending June 19, 2016

The crew closed the hatches of the Cygnus resupply ship on June 13. The following day, while the station was over Paraguay, Kopra, controlling the Canadarm2, they released the Cygnus spacecraft at 9:30 a.m. EDT. After Cygnus was a safe distance away, ground controllers at Glenn Research Center in Cleveland, Ohio, initiated the sequence for Saffire-1, and controllers at Orbital ATK in Dulles, Virginia, activated the experiment. This was an experiment to video fire in zero gravity and see how it would react during re-entry.

Malenchenko, Kopra and Peake continued to pack their Soyuz on June 15 and Williams prepared U. S. spacesuits for upcoming spacewalks. He sampled the cooling loop water then scrubbed the cooling loops. Skripochka loaded cargo and rubbish into the Progress 63 and Ovchinin used the Plasma Kristall experiment investigating charged micro-particles in plasmas.

At 9:20 a.m. EDT on June 17, Jeff Williams assumed command of the International Space Station from Tim Kopra. At 10:34 p.m. EDT, the Soyuz hatch closed between the station and the Soyuz TMA-19M spacecraft containing Kopra, Peake and Malenchenko. After spending 186 days aboard the ISS, they undocked from the station at 1:52 a.m. EDT, 254 miles over eastern Mongolia, and landed at 09:14 GMT in the steppes of Kazakhstan.

Having completed his sixth mission, Malenchenko now had spent 828 days in space, making him second on the all-time list behind Russian cosmonaut Gennady Padalka. Kopra now has 244 days in space on two flights while Peake spent 186 days in space on his first.

Postscript

At the time of writing, Principia was Timothy Peake's last spaceflight.

Tim works at ESA's Astronaut Centre in Cologne, Germany, as Astronaut Operations Team Lead. In June 2019, he announced that he will take an unpaid leave of absence from ESA for 2 years from October 1, 2019. During this time, he will return to the UK, where he will work more closely with the United Kingdom Space Agency (UKSA) on their education and outreach program. He will continue in his role as a Science, Technology, Engineering and Mathematics (STEM) ambassador for the UK, in addition to his outreach work as a Prince's Trust and Scout ambassador.

Fig. 7.13. Tim Peake with *Apollo 15* command module pilot Al Worden, Farnborough International Airshow, July 2018 (NASA)

During his single spaceflight, Peake logged 185 days, 22 hours and 11 minutes in space.

8

Proxima

Mission

ESA Mission Name:	Proxima
Astronaut:	Thomas Gautier Pesquet
Mission Duration:	196 days, 17 hours, 50 minutes, 18 seconds[1]
Mission Sponsors:	ESA/CNES
ISS Milestones:	ISS 49S, 86th crewed mission to the ISS

Launch

Launch Date/Time:	November 17, 2016, 20:20 UTC
Launch Site:	Pad 1, Baikonur Cosmodrome, Kazakhstan
Launch Vehicle:	Soyuz MS
Launch Mission:	Soyuz MS-03
Launch Vehicle Crew:	Oleg Viktorovich Novitsky (RKA), CDR
	Thomas Gautier Pesquet (ESA), Flight Engineer
	Peggy Annette Whitson (NASA), Flight Engineer

Docking

Soyuz MS-03

Docking Date/Time:	November 19, 2016, 21:58 UTC
Undocking Date/Time:	2 June 2017, 10:47 UTC
Docking Port:	Rassvet nadir

---

[1] Including the December 31, 2016 leap second.

© Springer Nature Switzerland AG 2020
M. von Ehrenfried, *European Missions to the International Space Station*,
Springer Praxis Books, https://doi.org/10.1007/978-3-030-38513-2_8

Landing

Landing Date/Time:	June 2, 2017, 14:10 UTC
Landing Site:	Near Dzhezkazgan, Kazakhstan
Landing Vehicle:	Soyuz MS
Landing Mission:	Soyuz MS-03
Landing Vehicle Crew:	Oleg Viktorovich Novitsky (RKA), CDR
	Thomas Gautier Pesquet (ESA), Flight Engineer

ISS Expeditions

ISS Expedition:	Expedition 50
ISS Crew:	Robert Shane Kimbrough (NASA), ISS-CDR
	Andrei Ivanovich Borisenko (RKA), ISS-Flight Engineer 1
	Sergei Nikolaevich Ryzhikov (RKA), ISS-Flight Engineer 2
	Oleg Viktorovich Novitsky (RKA), ISS-Flight Engineer 3
	Thomas Gautier Pesquet (ESA), ISS-Flight Engineer 4
	Peggy Annette Whitson (NASA), ISS-Flight Engineer 5

ISS Expedition:	Expedition 51
ISS Crew:	Peggy Annette Whitson (NASA), ISS-CDR
	Oleg Viktorovich Novitsky (RKA), ISS-Flight Engineer 1
	Thomas Gautier Pesquet (ESA), ISS-Flight Engineer 2
	Fyodor Nikolayevich Yurchikhin (RKA), ISS-Flight Engineer3
	Jack David Fischer (NASA), ISS-Flight Engineer 4

The ISS Story So Far

In March 2016, Soyuz TMA-20M delivered three crewmembers to the join Expeditions 47 and 48. In April, the Bigelow Expandable Activity Module (BEAM) was delivered by the SpaceX CRS-8 Dragon spacecraft and temporarily mounted on the Tranquility module for a feasibility study of the role of inflatable modules for future space station designs.[2]

A new version of the Soyuz spacecraft, Soyuz MS, with updated communications and navigation equipment, was launched for the first time in July 2016. Soyuz MS-01 delivered Expedition 48 and 49 crew to the station.

---

[2]The intention was to jettison BEAM after a year, but in October 2017 NASA announced that the module would stay in place until 2020, with options for two further 1-year extensions.

Shenzhou 11 was launched in November 2016 with Jing Haiping and Chen Dong for China's second space station, Tiangong-2, and their 33 days mission was the longest for that nation to-date.

Soyuz MS-02 delivered three crewmembers to the ISS in October 2016.

Thomas Pesquet

Thomas Pesquet was born in Rouen, France, on February 27, 1978. He was educated at the Lycée Pierre Corneille in Rouen, France, graduating in 1998. He went on to study at the École Nationale Supérieure de l'Aéronautique et de l'Espace in Toulouse, France, majoring in spacecraft design and control. He spent his final year before graduation at the École Polytechnique de Montréal, Canada, as an exchange student enrolled in the Aeronautics and Space Master course. He received his master's degree in 2001.

After university he worked as a trainee engineer with Thales Alenia Space in Cannes, France, developing a satellite system design tool and at GMV S.A., Madrid, Spain, as a spacecraft dynamics engineer on remote sensing missions. From 2002 to 2004, he worked at the French space agency, CNES, as a research engineer on space mission autonomy, future European ground segment design and European space technology harmonization. From late 2002, he was a CNES representative at the Consultative Committee for Space Data Systems, working on cross-support between international space agencies.

Pesquet was accepted onto Air France's flight training program in 2004 and he flew the Airbus A320 from 2006. Having logged more than 2300 flight hours on commercial airliners, he became a type rating instructor on the A320 and a Crew Resource Management instructor.

He was selected by ESA in May 2009 and joined the ESA astronaut corps in September 2009. He completed basic training in November 2010. After graduation, he worked as a Eurocom, communicating with astronauts during spaceflights from the mission control center. He was also in charge of future projects at the European Astronaut Centre, including initiating cooperation with new partners such as China. He has completed training on the Soyuz, U. S, and Russian spacesuits and on ISS systems. In 2011, he participated in the first CAVES training course, along with fellow ESA classmate Tim Peake. In September 2013 he was a member of the NEEMO 17 crew, also known as Seatest-2, along with Andreas Mogensen (ESA), Soichi Noguchi (JAXA) and Joseph Acaba and Kate Rubins (NASA). He returned to the Aquarius laboratory in July 2014 as part of the NEEMO 18 crew. He served as backup to Andreas Mogensen during the preparation for the Iriss mission, which launched in September 2015.

Pesquet is married to Anne Mottet and enjoys judo, basketball, jogging, swimming and squash as well as mountain biking, kite surfing, sailing, skiing, mountaineering, scuba diving and parachuting. He speaks French, English, Spanish, German and Russian.

Fig. 8.1. Thomas Pesquet (www.spacefacts.de)

He is a member of the following organizations:

- French Aeronautics and Astronautics Association (3AF)
- American Institute of Aeronautics and Astronautics (AIAA)

The Proxima Mission

Mission Patches

The name Proxima was chosen from over 1300 entries to an ESA competition. The winner was 13-year-old Samuel Planas from Toulouse, France. He said "Proxima is the closest star to our Sun and is the most logical first destination for a voyage beyond our Solar System. Proxima also refers to how human spaceflight is close to people on Earth." This name continues the tradition of French astronauts naming their missions after stars and constellations.[3]

ESA explains that "the logo continues the exploration theme, with star trails evoking future space travel and exploration beyond low-Earth orbit. Two stylized planets can represent our Earth and Moon or the Moon and Mars. The 'x' in Proxima is centred in the middle of the patch to signify the star Proxima Centauri. It also refers to the unknown, as well as Thomas being the tenth French space voyager. The three colored vertical lines form an outline of the International Space Station and also represent Earth, the Moon and Mars, as well as hinting at the French national flag". The patch was designed by Thomas Pesquet and Karen Oldenburg. Thomas commented: "I am really pleased with this mission name and the logo. It ticks all the boxes I had in mind by continuing the naming tradition for French astronauts and recognizing the legacy of human spaceflight so far while also being forward-looking and futuristic."

The Soyuz MS-03 crew patch was designed by Luc van den Abeelen featuring a classical 'shield' design. The flags represent the countries of the astronauts and the patch is topped with the Roscosmos logo. Three animals symbolize each of the crew members occupying three quadrants, while a fourth quadrant depicts the Soyuz spaceship, shown flying towards the space station's docking target. The eagle is taken from the state seal of Iowa, the birth state of U. S. astronaut Peggy Whitson. The Zubr buffalo from Belarus represents the Russian commander's origins, while the lion represents Thomas's origins from Normandy in France. Behind the crew's family names, a mountain in the Caucasus Mountains is shown – Kazbek, which is also the callsign for the crew.

---

[3] Michel Tognini's Antares mission to Mir on board Soyuz TM-15 in July 1992. Jean-Pierre Haigneré's Altair mission to Mir on board Soyuz TM-17 in July 1993 and his Perseus mission to Mir on board Soyuz TM-29 in February 1999. Claudie Haigneré's Cassiopée mission to Mir on board Soyuz TM-24 in August 1996 and her Androméde mission to the ISS on board Soyuz TM-33 in October 2001. Léopold Eyhart's Pégase mission to Mir on board Soyuz TM-27, January 1998.

Fig. 8.2. Proxima mission patch (ESA)

Fig. 8.3. Soyuz MS-03 mission patch (www.spacefacts.de)

Fig. 8.4. Expedition 50 mission patch (NASA)

Fig. 8.5. Expedition 51 mission patch (NASA)

The Expedition 50 patch is a simple rectangle with the crew names in the red border and the stylized number 50 hosting Earth and the Moon. The ISS orbits between the two, and there are six stars to represent the station crew.

The Expedition 51 patch was described by commander Peggy Whitson "From as early as the 11th century, coats of arms have been used as emblems representing groups as small as families to as large as countries. The Expedition 51 patch is designed as a modernized international coat of arms, blending the traditional shield shape with our modernized symbol of achievement, the International Space Station (ISS). The background represents our home world and its inhabitants, and on the right, outer space. The bi-color ISS is the bridge between the two, symbolizing the benefits on Earth of space research, and at the same time our mission to explore deeper into space, on a path to further discovery and knowledge."

Fig. 8.6. The Soyuz MS-03 crew with Thomas Pesquet on right (www.spacefacts.de)

Timeline

Expedition 50, Week Ending November 20, 2016

Thomas Pesquet of ESA, Peggy Whitson of NASA and Roscosmos commander Oleg Novitsky were launched in their Soyuz MS-03 spacecraft from the Baikonur cosmodrome in Kazakhstan at 20:20 GMT on November 17. This was Pesquet's first spaceflight and trip to the ISS, Whitson's third and Novitsky's

Fig. 8.7. The Expedition 50 crew with Thomas Pesquet third from the right (NASA)

Fig. 8.8. The Expedition 51 crew with Thomas Pesquet third from the right (NASA)

second. After orbiting Earth for approximately 2 days, their Soyuz MS-03 spacecraft docked with the space station's Rassvet module at 4:58 p.m. EST on November 19. The hatches were opened at 7:40 p.m. EST, and the new crew members were welcomed by Expedition 50 commander Shane Kimbrough and cosmonauts Sergey Ryzhikov and Andrey Borisenko.

Expedition 50, Week Ending November 27, 2016

Getting straight to work on November 21, Kimbrough and Pesquet commanded the Canadarm2 to release the Cygnus spacecraft at 8:22 a.m. EST while the space station was flying 251 miles over the Pacific Ocean, off the west coast of Colombia. The second Saffire-II experiment was initiated on board the Cygnus, by controllers at Glenn Research Center in Cleveland, Ohio and at Orbital ATK in Dulles, Virginia, once the craft was safely clear of the station. This experiment investigated flames in zero gravity with a view to designing safer spacecraft.

Whitson, Pesquet and Novitsky familiarized themselves with station emergency equipment and procedures. Whitson and Novitsky caught up with changes and additions to the station since their last visits.

Expedition 50, Week Ending December 4, 2016

Pesquet carried on from Tim Peake's research, and on November 29 used the Muscle Atrophy Research and Exercise System (MARES) chair. He investigated the muscles of his calf and his Achilles tendon. These muscles are not doing the work in space that they do on Earth supporting the weight of the body. He also carried out some tests of his ankles.

Progress 65 launched from the Baikonur cosmodrome in Kazakhstan at 9:51 a.m. EST on December 1. It carried more than 2.6 tons of food, fuel and supplies.

In general, the crew investigated fire and heat in weightlessness. They took samples from the Electrostatic Levitation Furnace, heated then and measured their thermophysical properties at high temperatures. They found a blockage in the Combustion Integrated Rack, so delayed some modifications until further investigation as carried out.

Expedition 50, Week Ending December 11, 2016

Kimbrough and Pesquet prepared for the arrival of the Japanese HTV-6 cargo ship. On December 5, they simulated the grapple and capture using the Canandarm2 robotic arm. Kimbrough scrubbed the cooling loops on U. S. spacesuits. Pesquet and Whitson conducted ultrasound scans as part of the Cardio Ox study. This looks at the long-term risk of atherosclerosis in station crew. On December 7,

Pesquet conducted and ultrasound scan on his neck, thighs and heart. He also wore the "smart shirt" as part of the EVERYWEAR study. The shirt has sensors to record biometric data.

The Japanese H-IIB rocket launched at 8:26 a.m. EST on December 9 from the Tanegashima Space Center in southern Japan. At the time of launch, the space station was flying about 250 miles over the Philippine Sea south of Japan. The HTV-6 cargo spacecraft successfully separated from the rocket at 8:41 a.m. EST and started its 4-day journey to the ISS.

Expedition 50, Week Ending December 18, 2016

Ground controllers successfully berthed the JAXA Kounotori 6 H-II Transfer Vehicle (HTV-6) at the Earth-facing port of the Harmony module at 8:57 a.m. EST on December 13. It contained more than 4.5 tons of supplies, water, spare parts and experiment hardware. While the crew slept robot controllers at mission control extracted a pallet of new batteries HTV-6 unpressurized area using the Canadarm2. The batteries were destined for the starboard truss. First thing in the morning, the crew opened the hatches to the HTV-6 resupply vehicle and began unloading.

Kimbrough then began the installation of a new Japanese experiment on space radiation exposure. Pesquet analyzed water samples looking for microbes. Whitson used the new touch-based tablet devices. Novitsky studied plasma physics and tested remote rover commands. Ryzhikov collected radiation detectors for the Matryeshka-Bubble experiment.

On December 15, Kimbrough unloaded the HTV-6. Whitson set up a replacement nano-satellite deployer in the Kibō laboratory. Kimbrough spoke with students at the Nantucket New School, in Massachusetts. Pesquet studied the AquaMembrane water recycling system and helped Kimbrough and Whitson conduct retina tests for the Ocular Health study. Ryzhikov and Borisenko carried out routine maintenance in the Russian segment, particularly on communication hardware.

Pesquet wrote on his Facebook page on December 16: "We have started well in advance to prepare for the spacewalks of January. It is a lot of work to service the suits and get them ready, get familiar with the choreography and prepare the tools and equipment. Not even mentioning the thousands of hours of work for all the personnel on the ground."

Expedition 50, Week Ending December 25, 2016

On December 20, Ryzhikov helped Pesquet measure his muscle and tendon responses wearing electrode stimulators his right calf as part of the Sarcolab experiment. Whitson replaced racks in the Fluids Integrated Rack and the

Combustion Integrated Rack. The former studies fluids and the latter investigates flames and burning materials. Kimbrough replaced cartridges in the Electrostatic Levitation Furnace, which studies the levitation, melting and solidification materials. He later joined Ryzhikov and Borisenko for an emergency Soyuz descent drill. They simulated an emergency escape and evacuation using their Soyuz spacecraft.

The crew installed the new small-satellite deployer the Kibō laboratory to replace the troublesome one, on December 22. It can hold twice as many satellites as its predecessor.

Later in the week, Kimbrough, Whitson and Pesquet prepared for a pair of spacewalks, scheduled for January 6 and 13, to complete the replacement of old nickel-hydrogen batteries with new lithium-ion batteries on the station's truss structure.

Expedition 50, Week Ending January 1, 2017

The Expedition 50 crew spent the last week of the year on a variety of science and maintenance tasks. They also prepared for upcoming EVAs. Kimbrough, Whitson and Pesquet continued their preparations for their pair of spacewalks to replace old nickel-hydrogen batteries with new lithium-ion batteries on the station's truss structure. They scrubbed the cooling loops on the spacesuits, reviewed EVA operations and did a suit fit check.

They participated in the Fluid Shifts study, which investigates the causes for lasting physical changes to astronauts' eyes; performed the final harvest of the Romaine Lettuce from the Veggie facility and continued preparing the station's Combustion Integrated Rack (CIR) for the upcoming Cool Flames Investigation. They enjoyed a usual weekend off and an extra day off on January 2.

Expedition 50, Week Ending January 8, 2017

Before the spacewalk on Friday, ground controllers used the Canadarm2 and Dextre to install three new lithium-ion batteries in the ISS's 3A power channel Integrated Electronics Assembly (IEA) pallet on the starboard 4 truss. Three of the old nickel-hydrogen batteries were stored on the Japanese H-II Transfer Vehicle's external pallet for return to Earth while the last one was temporarily stored on a platform on Dextre.

Pesquet and Borisenko configured the SPHERES for a schools competition. Ryzhikov and Novitsky carried out routine maintenance on Russian life support systems.

Kimbrough and Whitson Two started their spacewalk at 7:23 a.m. EST on January 6. They installed adapter plates and installed the power cables for three new lithium-ion batteries on the starboard truss. Ultimately, nine old nickel-hydrogen batteries would be placed in the external pallet of the HTV destined to burn up on re-entry. The spacewalk ended at 1:55 p.m. EST. During the 6 hours and 32 minutes spacewalk, they also accomplished several get-ahead tasks, including a photo survey of the Alpha Magnetic Spectrometer (AMS).

Expedition 50, Week Ending January 15, 2017

The entire crew had a day off on January 9, after the busy day of spacewalking the previous week and another planned for this week. In preparation for the second EVA, Kimbrough, Pesquet and Whitson reviewed procedures. Ground controllers used the Dextre to move the final lithium-ion battery to the 1A power channel Integrated Electronics Assembly, moved another nickel-hydrogen battery to one of Dextre's arms for temporary stowage and tightened down bolts on two of the previously moved Li-ion batteries.

Kimbrough and Pesquet switched their spacesuits to battery power and started the second of two spacewalks at 6:22 a.m. EST on January 13. The new channel lithium-ion batteries installed the previous week were working as planned and power was routed nominally. They installed three new adapter plates and connected three of the six new lithium-ion batteries. They concluded their spacewalk at 12:20 p.m. EST after carrying out some get-ahead tasks. They stowed Node-3 padded shields and they photographed equipment to be used on the next EVAs. This was the 197th spacewalk in support of assembly and maintenance of the orbiting laboratory.

ESA's Luca Parmitano, an experienced spacewalker himself, was lead communicator at mission control in Houston. Pesquet said: "I have so much respect for all the teams who designed, built, launched the hardware and planned, tested, conceived, conducted the spacewalks. And all the teams who trained us so that we did not make too many mistakes! My hat off to everybody."

Expedition 50, Week Ending January 22, 2017

The crew returned their focus to scientific research after the recent spacewalks on January 18. Pesquet repressurized the Japanese Kibō laboratory airlock after a small satellite deployer launched some cubesats. Whitson joined Borisenko and Novitsky for the ongoing Fluid Shifts study and collected blood, urine and saliva samples. Kimbrough worked on combustion science equipment troubleshooting a pair of devices that explore flames and high temperatures in space. Ryzhikov recharged Soyuz computer batteries.

The following day with Borisenko wearing the Lower Body Negative Pressure (LBNP) suit, Ryzhikov and Pesquet took an ultrasound scan and eye tests on Borisenko to test the effectiveness of the suit. The suit draws fluids down to the feet. Whitson set up vacuum access ports in the Harmony module. Kimbrough connected gas and water umbilical hoses in the Columbus laboratory module.

Kimbrough and Pesquet attempted to optimize the stations ability to exchange data between different science racks.

Expedition 50, Week Ending January 29, 2017

On January 24, teachers and students from Ireland, Romania and Portugal connected by video with Pesquet. Students and teachers gathered in Limerick, Timişoara and Lisbon, respectively, for the event organized by local European Space Education Resource Office (ESERO) offices. ESERO designs and disseminates classroom resources, tailored to the national school curricula and language – which uses space to make teaching and learning of STEM subjects more appealing and effective.

After ground operators unberthed the HTV on January 27, Pesquet and Kimbrough operated the Canadarm2 robotic arm to release it from the station at 10:46 a.m. EST. At the time of release, the station was flying 261 statute miles above the southern Atlantic Ocean.

Expedition 50, Week Ending February 5, 2017

Whitson and Pesquet participated in the Fine Motor Skills experiment, using an Apple iPad to investigate how the crewmembers interact with new technologies. Kimbrough rebooted a laptop on the MERLIN science freezer and replaced hard drives on a meteor recorder. He recorded himself reading of children's book for schools. Novitsky, Borisenko and Ryzhikov investigated bone loss and digestion in space and carried out routine maintenance in the Russian segment.

Progress 64 undocked from the Pirs docking compartment at 9:25 a.m. EST on January 31. On February 1, Whitson and Kimbrough welcomed attendees at the Johnson Space Center to Super Bowl festivities. Houston would host Super Bowl LI on the following Sunday.

On February 2, Whitson entered the BEAM module for a short time to install temporary sensors and literally punch the walls to measure the impacts. The vibrations indicate how the module copes with collisions. Pesquet wore the SkinSuit and documented his comfort, range of motion and other aspects of the suit.

February 5, JAXA's Kounotori, HTV-6 resupply ship fired its engines, deorbited and burned up over the southern Pacific.

Expedition 50, Week Ending February 12, 2017

Kimbrough, Whitson and Pesquet talked to ground planners on Monday, to review the mission profile, training materials and rendezvous procedures for the upcoming Dragon CRS-10 arrival. Pesquet prepared the airlock in the Kibō laboratory for the installation of a high-definition video camera mounted outside for recording the Earth's surface.

The crew trained for the robotic capture of the SpaceX Dragon and conducted an emergency Soyuz descent drill.

On February 9, Pesquet and Whitson replaced a gas bottle in the Combustion Integrated Rack in the Destiny laboratory. Later, Pesquet spent a day in the Kibō module working on science equipment maintenance. Whitson installed a leak locator in Kibō's airlock to locate the source of an ammonia leak outside the module.

Expedition 50, Week Ending February 19, 2017

Borisenko installed rodent habitats to investigate bone and tissue loss in weightlessness. Whitson installed new equipment in the Microgravity Science Glovebox to cultivate human stem cells. The station's guidance, navigation and control systems and its command and control systems were updated with new software transferred from the ground.

A pair of Synchronized Position Hold, Engage, Reorient, Experimental Satellites (SPHERES) were deployed inside the Kibō module to test new algorithms and docking techniques. Kimbrough checked out SpaceX communication equipment in readiness for the upcoming SpaceX CRS-10 mission. He then joined Whitson and Pesquet for periodic eye exams. They also collected blood and urine samples and stowed them in a science freezer.

Later in the week, Kimbrough and Pesquet practiced capturing the Dragon cargo ship using the Canadarm2 robotic arm. Whitson investigated intermittent power failures that were occurring on the Robonaut.

Expedition 50, Week Ending February 26, 2017

A busy week for new uncrewed cargo ships arriving at the station started with the launch of the Falcon 9 rocket at 9:39 a.m. EST on February 19. It carried the Dragon CRS-10 cargo ship. On February 22, the Russian Progress 66 launched at 12:58 a.m. from Baikonur, Kazakhstan. This was the first launch of a Progress cargo ship since the Progress 65 supply which failed to reach orbit on 1 December 2016. On the same day, the SpaceX Dragon rendezvous was postponed at 3:25 a.m. EST when onboard computers triggered the abort after recognizing an incorrect value in navigational data about the location of Dragon relative to the space station. The spacecraft was safe and was sent on a "racetrack" trajectory in front of, above and behind the station.

The second attempt was successful as Kimbrough and Pesquet captured Dragon at 5:44 a.m. EST on February 23. The craft was berthed to the Harmony module of the International Space Station at 8:12 a.m. EST. Finally, on February 24, the Progress 66 Russian cargo ship docked at 3:30 a.m. EST to the Pirs Docking compartment.

Expedition 50, Week Ending March 5, 2017

The crew housed the rodents delivered by Dragon in the Rodent Research-4 study. Stem cells were stored in a freezer. Whitson examined the stem cells under microscope. Plants were put in the Veggie facility. High-quality crystal samples were examined as part of the Light Microscopy Module Biophysics-1 experiment.

On February 28, Pesquet scrubbed the cooling loops and collected water samples from the U. S. spacesuits. Then he charged batteries on the helmet lights and tools. Kimbrough and Whitson maintained various life support equipment including replacing carbon dioxide filters in the Destiny laboratory. Borisenko studied how viruses behave in space while Ryzhikov and Novitsky explored non-invasive ways to monitor health.

Ground controllers switched from one pump to another in the thermal cooling system for one of the particle detectors on the Alpha Magnetic Spectrometer (AMS-02) on March 1. This is an experiment to study cosmic ray particle physics from the outside of the ISS. AMS was launched in 2011 with ESA's Roberto Vittori on board STS-134. At a cost of $2 billion, it is the most expensive experiment to date launched into space.

The Zvezda module fired its main engines for 43 seconds on March 2, raising the station's orbit to facilitate the departure and arrival of crew.

Pesquet began an 11-day prescribed diet, during which he would collect urine samples and measure his breathing for the energy study. Results will help researchers plan meals to ensure successful missions farther out into space.

Expedition 50, Week Ending March 12, 2017

Four cubesats were ejected on March 6 from the Kibō module's NanoRacks cubesat deployer. The LEMUR-2 satellites help monitor global ship tracking and improve weather forecasting. Pesquet joined Ryzhikov to conduct ultrasound scans of the head and neck for the long-running Fluid Shifts experiment. The following day, Pesquet took urine samples and measured Whitson's body to influence future space suit design.

Expedition 50, Week Ending March 19, 2017

On March 19, Pesquet released the SpaceX Dragon cargo spacecraft from the Canadarm2 robotic arm at 5:11 a.m. EDT. The Dragon cargo craft splashed down in the Pacific Ocean at 10:46 a.m. EDT, about 200 miles southwest of Long Beach, California.

Expedition 50, Week Ending March 26, 2017

Kimbrough and Pesquet reviewed procedures for their upcoming spacewalk. Pesquet talked with the French President Francois Hollande, who was with students near Lyon, France.

Pesquet opened the hatches to the BEAM on 22 March for a status check. He sampled air and surfaces for microbes and installed impact sensors. He also used a digital camera with a fish-eye lens to capture 360-degree imagery of the inside of BEAM.

Borisenko and Novitsky collected their blood, saliva and urine samples for a metabolism study and an immunity experiment.

Kimbrough and Pesquet switched their spacesuits to battery power at 7:24 a.m. EDT on March 24, starting their spacewalk.

Fig. 8.9. Shane Kimbrough (center), Oleg Novitsky (left) and Thomas Pesquet (right) in the airlock, March 2017 (ESA)

They disconnected the electrical connections of the Pressurized Mating Adapter-3 in advance of its relocation later that week. The International Docking Adapter will be attached to the PMA-3 for future commercial crew vehicle dockings. They lubricated the latching end effector on the Special Purpose Dexterous Manipulator, inspected a radiator valve and replaced cameras on the exterior of the Kibō. On Sunday, ground controllers used the Canadarm2 to relocate the PMA-3 from the Tranquility module to the Harmony module.

Expedition 50, Week Ending April 2, 2017

Whitson and Pesquet took more measurements of various body parts to investigate how life in space affect body size and shape. Whitson set up the Advanced Colloids Experiment Temperature Control-1 (ACE-T-1) physics experiment. She opened the Fluids Integrated Rack and reconfigured the Light Microscopy Module to investigate tiny suspended particles designed by scientists and observe how they form organized structures within water.

The second successive spacewalk started on March 30. Kimbrough and Whitson installed an upgraded computer relay and connected electrical connections on the Pressurized Mating Adapter-3. They installed four thermal protection shields on the Tranquility module where there was a gap after the PMA-3 was moved. The pair lost one of the shields but they were able to reuse a shield that they had removed previously. The EVA was concluded at 2:33 p.m. EDT. This was the 199th space-walk in support of assembly and maintenance of the orbiting laboratory.

At 11:51 a.m. EDT, Peggy Whitson broke the record for cumulative spacewalking time by a female astronaut. With over 50 hours and 40 minutes recorded, she passed the record of NASA astronaut Sunita Williams.[4] This was Whitson's eighth spacewalk.

Expedition 50, Week Ending April 9, 2017

Kimbrough tested the cooling water in the spacesuits they wore on the recent EVA. Whitson tested the SPHERES to demonstrate autonomous docking manoeuvres. Ryzhikov and Borisenko continued packing the Soyuz MS-02 spacecraft in anticipation of their homeward journey.

On April 9, Kimbrough handed over the command of station to Peggy Whitson. This was her second time commanding the ISS.[5]

Expedition 50/51, Week Ending April 16, 2017

At 12:45 a.m. EDT on April 10, hatches were closed between the Poisk module and the Soyuz MS-02 spacecraft. Expedition 50 crew members Shane Kimbrough of NASA and Sergey Ryzhikov and Andrey Borisenko of Roscosmos undocked at 3:57 a.m. The deorbit burn was at 6:28 a.m., and the crew landed southeast of Dzhezkazgan in Kazakhstan.

---

[4] Sunita Williams' last spacewalk was on November 1, 2012, with Akihiko Hoshide during Expeditions 33/34. Whitson went on to conduct two more EVAs (10 in total) and a reach a cumulative spacewalk time of 60 hours, 21 minutes – at the time of writing (July 2019) fourth place for all astronauts and still the record holder for female spacewalkers.

[5] Whitson previously commanded Expedition 16 in 2007/2008.

Expedition 51 began once Soyuz MS-02 undocked, with crew members commander Peggy Whitson of NASA and flight engineers Oleg Novitsky of Roscosmos and Thomas Pesquet of ESA.

Whitson examined the new interior station lighting and how it affected crew health and wellness. She continued to record her sleep patterns and took part in cognition and visual tests for the Lighting Effects study. Pesquet continued with the Fine Motor Skills test and then carried out spacesuit maintenance. He removed cooling water from the system and released built up gases from the suits' water tanks.

On April 13, Pesquet worked on the Electrostatic Levitation Furnace and studied the different phases of metallic alloys in the Material Science Research Rack. The next day, Whitson and Pesquet computer communications routers with help from Pesquet.

Expedition 51, Week Ending April 21, 2017

Whitson and Pesquet trained for the rendezvous and grapple of the upcoming Cygnus cargo ship. They practiced on a computer the robotic manoeuvres to capture Cygnus with the Canadarm2. The Cygnus CRS-7 or *John Glenn* was launched from Cape Canaveral SLC-41 on April 18 at 5:11 UTC.

Soyuz MS-04 launched from Baikonur, Kazakhstan at 3:13 a.m. EDT on April 20, when the station was flying about 250 miles over northeastern Kazakhstan near the southern Russian border. The craft carried two crew members, NASA's Jack Fischer and Fyodor Yurchikhin of Roscosmos. This was the first Soyuz MS to use the fast track 6 hours rendezvous. It was also the first Soyuz to carry only two crew members since Soyuz TMA-2.[6] This was the result of Russian budget cuts announced in 2016. The reduction of the Russian segment crew from three to two saved on Progress launches and freed up seats in Soyuz capsules. The Russian segment hasn't grown, but the USOS segment has, and the crew complement was to return to three once the Nauka Multi Purpose Module was launched and added to the Russian segment. However, this module has not launched at the time of writing, and each expedition from Expedition 51 to date has contained only two Russian cosmonauts.

The Soyuz MS-04 docked to the station's Poisk module at 9:18 a.m. EDT. The hatches were opened at 11:25 a.m. EDT and the pair were welcomed by Whitson, Novitsky and Pesquet.

---

[6] The ISS crew was reduced to a caretaker crew of two after the Columbia disaster in February 2003.

Pesquet and Whitson successfully captured the Cygnus cargo spacecraft *John Glenn* at 6:05 a.m. EDT on April 22. The cargo craft was berthed and bolted in place on the Earth-facing port of the Unity module at 8:39 a.m. EDT. It brought more than 3440 kilograms of research and supplies.

Expedition 51, Week Ending April 30, 2017

On April 24 Peggy Whitson broke the record for most cumulative time in space for an American. She equaled and surpassed Jeff Williams' record of 534 days set during his four spaceflights. She passed Williams' mark at 1:27 a.m. EDT. President Donald Trump called the ISS to congratulate Whitson on her record-breaking career.

The crew continued to unload cargo from both the spacecraft that arrived the previous week. Yurchikhin unloaded the Soyuz MS-04, while Whitson and Pesquet offloaded science experiments and crew supplies from the Cygnus.

ESA announced on April 24 that the crew broke another record in the previous month by conducting a combined 99 hours of science in the week beginning March 6.

Whitson and Fischer live-streamed using 4 K ultra-high-definition technology for the first time on April 26 when they talked to the National Association of Broadcasters in Las Vegas. Pesquet joined Novitsky for ultrasound scans and eye exams. He later joined Whitson and Fischer for more eye checks. Whitson also studied how astronauts adapt to touchscreen interfaces using an iPad. Fischer replaced sample cartridges in the high-temperature furnace laboratory facility.

Expedition 51, Week Ending May 7, 2017

Whitson and Pesquet started the process of the OsteoOmics bone experiment. This 4 weeks study, examined the molecular mechanisms that impact the bones of those living in space. Fischer and Whitson carried out an ultrasound scan of Fischer's leg muscles as part of the Sprint study. This investigates how high-intensity, low-volume exercise can maintain muscle, bone and heart functions. Whitson, Fischer and Yurchikhin had emergency training, taking note of safety equipment locations and escape paths to the docked Soyuz vehicles and making sure hatches had no obstructions. Pesquet recorded data for the student designed Genes in Space experiment. This looks at how DNA changes in space and how it affects the immune system.

Expedition 51, Week Ending May 14, 2017

Whitson continued looking at bone-loss and Fischer packed away leaves from the Veg-03 botany study.

Over the weekend, ground controllers used the Special Purpose Dexterous Manipulator (SPDM) arm to replace Main Bus Switching Unit-2 The old MBSU had stopped communicating on April 25 but was still routing power to station systems. The MBSU was bypassed during the replacement so station operations were not affected. This was the first time an MBSU was replaced using remote robotics and not by astronauts on a risky EVA.

On May 10, Whitson and Fischer reviewed procedures for the upcoming space-walk. Pesquet and Yurchikhin worked on a muscle study using electrodes attached to their legs while exercising.

The 200th spacewalk aboard the ISS began at 9:08 a.m. EDT on May 12. The spacewalk was shortened by planners, when the two astronauts ended up having to share a service and cooling umbilical (SCU) after a small water leak was detected in a second SCU. SCU's are not part of the suits but are used in pre-breathe activities in the Equipment Lock. Sharing the services reduced the battery power available for use during the spacewalk.

Whitson and Fischer ended the EVA at 1:21 p.m. EDT. They had replaced a large avionics box that supplies electricity and data connections to the science experiments. They also connected a data cable to the Alpha Magnetic Spectrometer, fixed insulation on the elbow of the Japanese robot arm, and installed a protective shield on the Pressurized Mating Adapter-3.

Expedition 51, Week Ending May 21, 2017

Fischer loaded the cubesat deployer in the Kibō module's airlock for the start of another series of cubesat launches. Six cubesats were deployed on May 16 to observe Earth's upper atmosphere and interstellar radiation left over from the Big Bang.

More cubesats were deployed on May 18, totaling over a dozen during the week. This batch of satellites had varied missions, including hybrid, low tempera-ture energy stowage systems and the upper reaches of Earth's atmosphere known as the thermosphere. After taking an ultrasound scan of his thigh and calf muscles, Fischer used the NeuroMapping experiment to see how the human brain structure and function changes in space. One of the station's multiplexer-demultiplexer (MDM) data relay boxes failed at 1:13 p.m. CET the day before. Station managers met Sunday make a plan. There are a pair of redundant MDM data relay boxes on the S0 truss so there was no impact on station operations.

Expedition 51, Week Ending May 28, 2017

Program managers gave the go ahead for a contingency spacewalk to replace the faulty MDM. Whitson and Kimbrough had upgraded the software of the failed MDM on a March 30 spacewalk.[7]

Whitson and Fischer successfully replaced MDM computer relay box and installed a pair of antennas on station to enhance wireless communication during their 2 hours and 46 minutes spacewalk on May 23.

Expedition 51, Week Ending June 4, 2017

Peggy Whitson handed over command of the International Space Station to Russian cosmonaut Fyodor Yurchikhin at 11:50 a.m. EDT on June 1. This is the first time that a commander handed over command to their successor and did not then leave the station. Due to the Russian's reducing their crew complement from three to two, there was a spare seat on the returning Soyuz MS-04, planned for a September landing. NASA and Roscosmos agreed to extend Whitson's tenure on board for a further 3 months.

The SpaceX Dragon launch was scrubbed due to lightning in the vicinity of the launch pad.

On June 2, Oleg Novitsky and Thomas Pesquet undocked their Soyuz MS-03 from the station at 6:47 a.m. EDT. After spending 196 days in space, they landed in Kazakhstan at approximately 10:10 a.m. EDT.

Postscript

After the Proxima mission, Pesquet returned to active duty in the ESA astronaut corps. His technical tasks include preparing for the future of human spaceflight including the lunar gateway. In 2018 he gained his Airbus A310 type rating and is qualified as a Novespace Zero-G aircraft pilot. In this role, he maintains his flight currency and flies the Airbus A310 parabolic aircraft for space agencies to run experiments in weightlessness.

In January 2019, ESA Director General Johann-Dietrich Wörner tweeted that he had proposed Pesquet for the next ESA flight to the ISS in late 2020 or early 2021. This would be after Luca Parmitano's Beyond mission in 2019.

---

[7] A similar MDM replacement spacewalk was conducted in April 2014 by Expedition 39 spacewalkers Steve Swanson and Rick Mastracchio.

Fig. 8.10. Thomas Pesquet with French President Emmanuel Macron and ESA Director General Jan Wörner, Paris Air and Space Show, June 2017 (ESA)

During his single spaceflight, Pesquet logged 196 days, 17 hours and 50 minutes in space.

9

Vita

Mission

ESA Mission Name:	Vita
Astronaut:	Paolo Angelo Nespoli
Mission Duration:	138 days, 16 hours, 56 minutes
Mission Sponsors:	ASI/ESA
ISS Milestones:	ISS 51S, 88th crewed mission to the ISS

Launch

Launch Date/Time:	July 28, 2017, 15:41 UTC
Launch Site:	Pad 1, Baikonur Cosmodrome, Kazakhstan
Launch Vehicle:	Soyuz MS
Launch Mission:	Soyuz MS-05
Launch Vehicle Crew:	Sergei Nikolayevich Ryazansky (RKA), CDR
	Randolph James Bresnik (NASA), Flight Engineer
	Paolo Angelo Nespoli (ESA), Flight Engineer

Docking

Soyuz MS-05

Docking Date/Time:	July 28, 2017, 21:54 UTC
Undocking Date/Time:	December 14, 2017, 05:14 UTC
Docking Port:	Rassvet nadir

Landing

Landing Date/Time:	December 14, 2017, 08:37 UTC
Landing Site:	near Dzhezkazgan, Kazakhstan
Landing Vehicle:	Soyuz MS
Landing Mission:	Soyuz MS-05
Landing Vehicle Crew:	Sergei Nikolayevich Ryazansky (RKA), CDR
	Randolph James Bresnik (NASA), Flight Engineer
	Paolo Angelo Nespoli (ESA), Flight Engineer

© Springer Nature Switzerland AG 2020
J. O'Sullivan, *European Missions to the International Space Station*,
Springer Praxis Books, https://doi.org/10.1007/978-3-030-30326-6_9

ISS Expeditions

ISS Expedition:	Expedition 52
ISS Crew:	Fyodor Nikolayevich Yurchikhin (RKA), ISS-CDR
	Jack David Fischer (NASA), ISS-Flight Engineer 1
	Peggy Annette Whitson (NASA), ISS-Flight Engineer 2
	Sergei Nikolayevich Ryazansky (RKA), ISS-Flight Engineer 3
	Randolph James Bresnik (NASA), ISS-Flight Engineer 4
	Paolo Angelo Nespoli (ESA), ISS-Flight Engineer 5
ISS Expedition:	Expedition 53
ISS Crew:	Randolph James Bresnik (NASA), ISS-CDR
	Sergei Nikolayevich Ryazansky (RKA), ISS-Flight Engineer 1
	Paolo Angelo Nespoli (ESA), ISS-Flight Engineer 2
	Aleksandr Aleksandrovich Misurkin (RKA), ISS-Flight Engineer 3
	Mark Thomas Vande Hei (NASA), ISS-Flight Engineer 4
	Joseph Michael Acaba (NASA), ISS-Flight Engineer 5

The ISS Story So Far

Soyuz MS-04 was the first Soyuz to carry only two crew members since Soyuz TMA-2 in 2002 due to a Russian decision to temporarily cut back their number of station crew numbers from three to two. Because there would now be a spare seat on the return journey, Whitson's mission was able to be extended by 3 months, and she returned to Earth together with Fyodor Yurichikhin and Jack Fischer.

Paolo Nespoli[1]

Paolo Nespoli was born on April 6, 1957, in Milan, Italy. He was drafted by the Italian army in 1977 and became a non-commissioned officer and parachute instructor at the Scuola Militare di Paracadutismo of Pisa. In 1980 he joined the ninth Btg d'Assalto 'Col Moschin' of Livorno, where he was a member of the Special Forces. From 1982 to 1984, he was assigned to the Italian contingent of the Multinational Peacekeeping Force in Beirut, Lebanon. On his return to Italy he was commissioned as an officer and remained with the Special Forces.

---

[1] The biography of Paulo Nespoli biography is taken from the previous book by the same author, *In the Footsteps of Columbus: European Missions to the International Space Station.* That book contains the details of his Esperia and Magisstra missions.

Fig. 9.1. Paolo Nespoli (NASA)

Paolo resumed his education in 1985, graduating from the Polytechnic University of New York in 1988 with a bachelor's degree in aerospace engineering and following this up in 1989 with a master's in aeronautics and astronautics. Having left the army in 1987, he returned to Italy to work as a design engineer for Proel Tecnologie in Florence, which manufactured ion propulsion units for satellites and spacecraft. There he conducted mechanical analysis and supported the qualification of the flight units of the Electron Gun Assembly, one of the main parts of the Italian Space Agency's Tethered Satellite System. He was awarded the Laurea in Ingegneria Meccanica by the Università degli Studi di Firenze in Italy in 1990.

In 1991 Paolo joined the European Astronaut Centre (EAC) in Cologne, Germany, as an astronaut training engineer. He contributed to basic training for European astronauts and was responsible for the preparation and management of astronaut proficiency maintenance, as well as the Astronaut Training Database – one of the systems used in the training process. In 1995 he worked on the EuroMir project at ESTEC in Noordwijk, the Netherlands, where he headed the team which prepared, integrated, and supported the Payload and Crew Support Computer that was used on the Mir space station. In 1996 he went to the NASA Johnson Space Center in Houston, Texas, and worked in the Spaceflight Training Division, training crews for the ISS.

In July 1998 Paolo joined the ASI astronaut corps and one month later he joined ESA's European astronaut corps. He was promptly assigned to Houston as a member of NASA's Astronaut Group 17, known as the Penguins, together with ESA astronauts Léopold Eyharts, Hans Schlegel, and Roberto Vittori. In 2000 Paolo qualified to fly on the Space Shuttle and to work on board the ISS. In July 2001 he completed the space shuttle robotics arm course and in September 2003 completed advanced skills training for spacewalks. In August 2004 he continued his training at the Gagarin Cosmonaut Training Centre in Star City, Moscow, learning to operate the Soyuz spacecraft.

After returning to the Johnson Space Center he was assigned to space shuttle mission STS-120 in June 2006.

In October 2007, Nespoli visited the station on board STS-120. This was his Esperia mission. He was a member of Expeditions 26 and 27 between December 2010 and May 2011 during his Magisstra mission.[2]

The Vita Mission

Mission Patches

ESA explains that "the mission name, Vita, stands for Vitality, Innovation, Technology and Ability and was chosen by the Italian space agency, ASI, which was providing the mission through a barter agreement with NASA. In Italian, 'vita' means 'life', reflecting the experiments that Paolo will run and the philosophical notion of living in outer space – one of the most inhospitable places for humans.

The overall circle and blue shading evoke our planet, with the Third Paradise symbol by Italian artist Michelangelo Pistoletto linking the mission's main messages. Three elements stand out: a strand of DNA as a symbol of life and science, a book as a symbol of culture and education, and Earth as a symbol of humanity. The Third Paradise is a reformulation of the symbol for infinity. The two opposing

---

[2] See the author's book, *In the Footsteps of Columbus, European Missions to the International Space Station.*

ovals contain elements of the scientific and cultural activities Paolo will perform in space. Their meeting in the center represents the evolution of Earth and benefits for humankind. The central shape of the symbol, together with the presence of the globe, can also be seen as an eye, giving an astronaut's perspective over our planet. The logo features the colors of the Italian flag".

The Soyuz MS-05 crew patch features a nose-on view of the spacecraft, as it prepares to dock with the International Space Station, with the Roscosmos logo just above it. In a protruding circle, the Greek god of the northern wind Boreas is shown, as this name is the call sign for the spaceship's crew. In the foreground, the navigational device of the docking cross is shown, as seen by the crew during the link-up with the orbital facility. The constellation of Scorpion is shown in the background, as a reference to the spaceship commander's star sign. Three stars in the background symbolize the crew members, whose names are shown in the border, with the corresponding national flags next to them. It was designed by Anastasia Timofeyeva, who won a competition with over 700 entries, and it was finalized by Luc van den Abeelen.

Fig. 9.2. Vita mission patch (ESA)

Fig. 9.3. Soyuz MS-05 mission patch (www.spacefacts.de)

Fig. 9.4. Expedition 52 mission patch (NASA)

Fig. 9.5. Expedition 53 mission patch (NASA)

The Expedition 52 patch is one of the busiest produced. It was designed by Luc van den Abeelen and Commander Fedor Yurchikin. It was originally to be the Expedition 53 patch but followed Expedition 52 as reassignments were made. As well the names and national flags of the crew in the border, it shows the ISS sharing an orbit of Earth with Sputnik. The roman numerals LII are highlighted in the solar arrays to indicate the mission number, 52. As well as stars of various colors and shapes, visible are the Moon, Mars, a comet, the Sun, Jupiter and Saturn. The stars can be seen to make simple shape of a house around the Earth to symbolise Yurchikin's mission slogan of "Earth is our home."

The Expedition 53 patch was designed by Tim Gagnon and Jorge Cartes. It shows the ISS on a flight path following the astronaut office symbol from Earth to the Moon and onwards past Mars. Earth is held in the number 5 and the Moon is in the number 3. Beyond both are stars, constellations and nebula.

Fig. 9.6. The Soyuz MS-05 crew with Paolo Nespoli on left (www.spacefacts.de)

Fig. 9.7. The Expedition 52 crew with Paolo Nespoli third from right (NASA)

Fig. 9.8. The Expedition 53 crew with Paolo Nespoli on right (NASA)

Timeline

Expedition 52, Week Ending July 30, 2017

On July 28, Soyuz MS-05 launched from Baikonur, Kazakhstan at 11:41 a.m. EDT. On board were NASA astronaut Randy Bresnik, Sergey Ryazanskiy of Roscosmos and Paolo Nespoli of ESA. They docked at the station's Rassvet module, at 6 p.m., and the hatches opened at 7:57 p.m. EDT. The new members of Expedition 52 were welcomed by Station Commander Fyodor Yurchikhin of Roscosmos and Flight Engineers Peggy Whitson and Jack Fischer of NASA. This was Nespoli's third spaceflight and his second long duration mission to the ISS.

Expedition 52, Week Ending August 6, 2017

Since the Russians reduced their crew from three to two, the USOS segment had increased its crew from three to four. The USOS segment is crewed by NASA, ESA, JAXA and CSA astronauts. Expedition 52 now had six members.

Nespoli took part in the Sarcolab-3 study using the Muscle Atrophy Research & Exercise System (MARES) chair in the Columbus module. This study continues the investigation of muscle loss, by recording data on the astronauts' calf muscles.

Nespoli and Bresnik participated in the research on space headaches by writing down their experiences on 2 August and for the remainder of the week. Ryazanskiy, Nespoli and Bresnik familiarized themselves with the station's emergency equipment and explored the station, taking note of safety equipment locations and escape paths.

Bresnik and Nespoli conducted ultrasound scans of their legs on August 3 to review how their leg muscles and tendons were changing in space, to be compared with pre- and post-flight data.

Expedition 52, Week Ending August 13, 2017

Bresnik and Nespoli started the week by returning to the muscle loss observations. Using electrodes and an ultrasound, they recorded muscle performance during exercise. Yurchikhin and Ryazanskiy prepared for their upcoming EVA.

The Russian Progress spacecraft fired its thrusters on August 9 to raise the orbit of the station to prepare for the following month's crew arrivals and departures. Fischer and Nespoli trained for the arrival and grappling of the next SpaceX Dragon. Yurchikhin and Ryazanskiy performed leak checks, installed batteries and sized up their Orlan spacesuits before the EVA.

Expedition 52, Week Ending August 20, 2017

SpaceX Dragon CRS-12 launched from Cape Canaveral Air Force Station at 16:31 UTC on August 14. Fischer and Nespoli captured the Dragon spacecraft at 6:52 a.m. EDT using the Canadarm2 robotic arm, and it was then berthed at the Harmony module at 9:07 a.m. EDT. CRS-12 delivered more than 6400 pounds of supplies and payloads to the station, including small cups of chocolate, vanilla and birthday cake-flavored ice cream in freezers. The freezers were to be reused to house scientific samples for the return journey.

Yurchikhin and Ryazansky started their spacewalk at 10:36 a.m. EDT on August 17 in their Orlan spacesuits. They exited the Pirs docking compartment. Ryazansky released five nanosatellites, including the first 3D printed cubesat into Earth orbit, by 'simply' throwing them in the right direction away from the Space Station into orbit. Other tasks included retrieving external experiments and taking samples of materials that are subjected to the harsh conditions of space to see how they fare over time. External experiments like these and the joint ESA-Russian Expose facilities allow mission designers to test potential new materials and even understand the origins of life.

Nespoli helped them prepare for the spacewalk and helped them back inside when they were finished. Fyodor was wearing a new version of the Orlan spacesuit that offers better temperature control and was easier to use and more robust.

Expedition 52, Week Ending August 27, 2017

Overnight on August 22, ground-based robotics controllers extracted a new astrophysics experiment from the unpressurized segment of the SpaceX Dragon, ready to be handed to the Japanese robotic arm to be installed outside the Kibō module. This was the Cosmic Ray Energetics and Mass Investigation (CREAM) experiment, which observes a variety of cosmic rays and measures their charges. The experiment is an extension of what started as high-altitude, long-duration balloon flights over Antarctica. The orbital data was expected to be several orders of magnitude greater than that collected in Earth's atmosphere.

The crew members watched solar eclipse from space on August 21. They used a multitude of cameras to photograph the eclipse.

Nespoli unloaded experimental equipment from the Dragon. These included the ESA incubator-centrifuge called Kubik, which combines a centrifuge with a heater element to keep samples at the right temperature. He plugged in the unit, ran tests and confirmed that Kubik was running at 37 °C. Also on board was the ASI experiment, SERISM, investigates on a molecular level why immune cells die more quickly in space. SERISM looks at human blood-derived stem cells from healthy donors on Earth that are transported to the Space Station in Experiment containers that slot into Kubik. Here they are reprogrammed to produce bone material in microgravity and preserved at different times using chemical fixatives for later analysis.

He also entered the BEAM inflatable module and took samples of areas to see if and how bacteria are growing on the new surfaces. In the afternoon he did a session with Aquapad to test the station's drinking water and more eye exams, as well as taking a fecal sample for the Japanese ProBiotics experiment investigating how bacteria in astronauts' gut changes during spaceflight.

Nespoli and Bresnik tested the effectiveness of the new Mini-Exercise Device-2 (MED2). The MED2 is smaller and less bulky than other space exercise equipment, providing more habitability room on a spacecraft.

Expedition 52, Week Ending September 3, 2017

Fyodor Yurchikhin handed over command of the station to Randy Bresnik on September 1 at 20:30 CEST in the usual formal ceremony. At 2:41 p.m. EDT on August 2, the hatch closed between the Soyuz and the station. Whitson, Fischer and Yurchikhin undocked their Soyuz MS-04 at 5:58 p.m. EDT and landed safely at 9:21 p.m. EDT southeast of Dzhezkazgan in Kazakhstan.

Expedition 53, Week Ending September 10, 2017

Nespoli checked out physics and life science equipment on September 6. He cleaned and installed handrails on the Electromagnetic Levitation device then swapped out equipment inside the Space Automated Bioproduct Lab. The following day, station commander Bresnik and Nespoli investigated the failed Main Bus Switching Unit (MBSU). They replaced electronics and connected it to a laptop. Nespoli set up the Magvector magnetic field experiment ahead of experimentation the following week. This studies the Earth's magnetic field and how it affects electrically powered experiments on the station. Nespoli also checked the Light Microscopy Module equipment that had to be verified after transport as it was fragile.

Expedition 53, Week Ending September 17, 2017

Soyuz MS-06 launched from Baikonur, Kazakhstan at 5:17 p.m. EDT, September 12 with Mark Vande Hei and Joe Acaba of NASA and Alexander Misurkin of Roscosmos on board. The Soyuz spacecraft docked to Poisk zenith port at 10:55 p.m. EDT while both spacecraft were flying 252 statute miles over the Pacific Ocean off to the west of Chile. Expedition 53 Commander Randy Bresnik of NASA and Flight Engineers Sergey Ryazanskiy of Roscosmos and Paolo Nespoli of ESA welcomed them on board the station.

Nespoli and Bresnik used the Canadarm2 robotic arm to release the SpaceX Dragon after it was detached from the Earth-facing port of the Harmony module at 4:40 a.m. EDT on September 17. Dragon's thrusters fired to move the spacecraft a safe distance from the station. Then SpaceX flight controllers in Hawthorne, California, commanded its deorbit burn. The capsule splashed down at 10:14 a.m. in the Pacific Ocean.

Expedition 53, Week Ending September 24, 2017

Nespoli and Ryazanskiy participated in the bone marrow study, looking at blood and breath samples with the blood being processed in a centrifuge. Bresnik also collected his blood and urine samples that scientists later analyzed for any physiological changes caused by microgravity. Nespoli assisted Ryazanskiy into the MARES chair and Bresnik collected ultrasound imagery of his leg. The data was readied for the Sarcolab-3 experiment that observes space-induced chemical and structural changes in muscle fibers.

On September 12 Bresnik and Vande Hei wore virtual reality headsets to train for emergency escape scenarios. They simulated using jet packs on their spacesuits in case they become untethered from the station.

Ryazanskiy and Nespoli continued the Sarcolab-3 study on muscle and tendon changes in weightlessness.

Expedition 53, Week Ending October 1, 2017

Bresnik and Nespoli worked out on the new Miniature Exercise Device-2 (MED-2). Vande Hei installed new lightbulbs in his crew quarters to help with circadian rhythms, sleep and cognitive performance. Acaba and Nespoli took blood and urine samples for the Biochemical Profile and Repository study which records changes to the body on long duration flights. Bresnik and Vande Hei prepared for an upcoming series of three spacewalks. They checked the spacesuit to station tethers and collected the required tools and equipment.

On September 26, Acaba set up the Veggie-3 experiment which grows lettuce and cabbage as a step in creating a sustainable food source in space. Bresnik and Vande Hei continued to check their emergency jet packs and they resized their spacesuits for the upcoming EVA.

Acaba installed radiation monitors in the U. S. segment on September 27 to record neutron radiation levels as part of the Radi-N2 study to examine the risk to crew members and develop advanced protection for future spacecraft.

Progress 67 fired its engines on September 28 for 3 minutes and 40 seconds to boost the station's orbit. This placed the station at the correct altitude to receive Progress 68 mid-October and Soyuz MS-07 in December.

Expedition 53, Week Ending October 8, 2017

Ground controllers moved the Canadarm2 to the correct worksite on October 2 to allow the spacewalkers access to its Latching End Effector (LEE).

Nespoli prepared the Kibō laboratory for new science equipment on October 3. These were due to arrive on Orbital ATK's Cygnus and SpaceX's Dragon, both in November.

Bresnik and Vande Hei started the first of a series of spacewalks at 8:05 a.m. EDT on October 5. They replaced one of the LEEs on the Canadarm2. They also accomplished a couple of get-ahead tasks such as removing insulation from a spare direct current switching unit and preparing a flex hose rotary coupler for future use.

Expedition 53, Week Ending October 15, 2017

Bresnik and Vande Hei switched their spacesuits to battery power again at 7:56 a.m., EDT on October 10 for their second successive spacewalk. This time they lubricated components of the new latching end effector they had installed on the Canadarm2 robotic arm the week before and replaced a faulty camera.

The planned launch of the Progress 68 cargo craft was scrubbed on October 12. It was rescheduled for October 14. It launched on schedule at 4:46 a.m. EDT Baikonur, Kazakhstan. At the time of launch, the ISS was flying about 250 miles over the southern Atlantic Ocean north of the Falkland Islands.

Expedition 53, Week Ending October 22, 2017

Progress 68 cargo ship docked at 7:04 a.m. EDT on October 16 at the Pirs Docking Compartment. Bresnik set up a camera that photographs meteors entering Earth's atmosphere. Acaba configured a microscope in the Fluids Integrated Rack and Nespoli set up the new Mini-Exercise Device-2 (MED-2).

Bresnik and Acaba reviewed procedures and configured tools on October 19, before their spacewalk. Vande Hei and Nespoli prepared to help the spacewalkers in and out of their spacesuits and guide the pair as they worked outside.

Bresnik and Acaba started their spacewalk at 7:47 a.m. EDT on October 20. They replaced a camera light assembly on the Canadarm2 latching end effectors (LEE) that the spacewalkers installed on the Canadarm2 on October 5. They also installed an HD camera outside the station. The EVA ended at 2:36 p.m. EDT. Some of the get-ahead tasks they completed include greasing the new end effector on the robotic arm, installing a new radiator grapple bar and Bresnik prepared a spare pump modules to make it accessible for future robotic replacement.

Expedition 53, Week Ending October 29, 2017

On October 23 the crew participated in the Lighting Effects experiment where LEDs replaced the fluorescent bulbs. The LEDs can be adjusted with choices of intensity and color. The effect of the lighting on circadian rhythms and sleep patterns were recorded by the crew.

At 11 a.m. EDT on October 25, Bresnik spoke with U. S. Ambassador to Ukraine, Marie Yovanovitch and Ukrainian and U. S. embassy students.[3] Nespoli took part in second day of an 11-day study called the Astronaut's Energy Requirements for Long-Term Space Flight (Energy). This examines the side effects of space travel. He collected water samples, urine samples and packed away the Pulmonary Function System equipment.

About 100 ml of freon leaked from a nanosatellite ready for launch from Kibō on October 27. There was no risk to crew health and safety nor to station hardware.

On October 26, the crew participated in a call with Pope Francis, who was in the Vatican. The Pope asked what motivated them to become astronauts/cosmonauts. Paolo Nespoli translated for the rest of the crew. From NASA blog October 26:

> *Nespoli indicated that while he remains perplexed at humankind's role, he feels their main objective is enriching the knowledge around us. The more*

---

[3] Ambassador Yovanovitch would later become embroiled in the "Ukraine scandal". She was removed from her post by Secretary of State Mike Pompeo and later testified before the House Intelligence Committee as part of the 2019 impeachment of President Donald Trump.

we know, the more we realize we don't. Part of the space station's ultimate mission is filling in those gaps and revealing the mysteries locked away in the cosmos. Ryazansky told the Pope that it was an honor to continue his grandfather's legacy aboard the orbiting laboratory. Ryazansky's grandfather was a chief engineer of Sputnik, the world's first satellite to launch to space. Ryazansky said he is now part of the future of humanity, helping to open frontiers of new technology. Bresnik spoke candidly to Pope Francis, saying that one cannot serve aboard the space station and not be touched to their soul. From Bresnik's unique vantage orbiting Earth, it is obvious there are no borders. Also evident: a fragile band of atmosphere protecting billions of people below. Pope Francis said that while society is individualistic, we need collaboration, and there is no better example of international teamwork and cohesiveness than the space station. It is the ultimate human experiment, showing that people from diverse backgrounds can band together to solve some of the most daunting problems facing the world. "The totality is greater than the sum of its parts," Pope Francis observed. At the end of the call, the Pope thanked his new friends, offered his blessings and asked that they, too, pray for him in return.

Expedition 53, Week Ending November 5, 2017

The crew harvested some fresh lettuce, cabbage and mizuna from the Veg-03 investigation. This investigates how to grow food on long duration missions and in the harsh environment of space and Mars. They also set up EarthKAM in Node 2 for a week-long imaging session.

The crew also spent part of November 1 photographing their "home life" onboard the ISS for the Canadian At Home in Space study. This investigation "assesses the culture, values and psychosocial adaptation of astronauts to a space environment shared by multinational crews during long mission time frames". The crew answer questionnaires to investigate if they, and other crews, develop a "unique, shared space culture as an adaptive strategy for handling the cultural differences they encounter in their isolated and confined environment by creating a home in space". Vande Hei and Acaba talked to pupils and teachers at Shaker Heights High School in Cleveland, Ohio.

On November 1, the crew stayed up late to watch Game 7 of the baseball World Series, which was won by the NASA astronauts' home team, the Houston Astros. They also won the series 4 to 3.

The station's altitude was raised during a 3 minutes, 26 seconds firing of Progress 67's thrusters.

On November 5 Nespoli captured a time-lapse video of a fireball falling to Earth over the Atlantic, off the South African west coast.

Expedition 53, Week Ending November 12, 2017

Vande Hei gathered saliva samples on November 6 as part of a study on the human immune system and metabolism. Bresnik took panoramic photographs inside the Kibō laboratory module in preparation for the Astrobee experiment. This consists of three free-flying, cube-shaped robots that should help the crew in their daily tasks and record their actions.

Fig. 9.9. Paolo Nespoli preparing to talk to Italian President Sergio Mattarella, November 2017 (ESA)

Nespoli wore a garment with water shielding the organs from radiation Vande Hei installed sensitive monitors to measure the crew exhaled air. Acaba replaced the hard drive and loaded software the Alpha Magnetic Spectrometer laptop. Bresnik fixed an audio speaker in the Harmony module.

The launch of Orbital ATK's Antares rocket and Cygnus cargo spacecraft was postponed on November 11 due to an aircraft infringing the no-fly zone. It lifted off at 7:19 a.m. EST on November 12.

Expedition 53, Week Ending November 19, 2017

At 5:04 a.m. on November 14, Nespoli and Bresnik successfully captured the Cygnus cargo spacecraft 'Gene Cernan'[4] using the Canadarm2 robotic arm. Robotic ground controllers berthed it at the Earth-facing port of the Unity module. The craft was bolted into place at 7:15 a.m. EST.

Bresnik carried out troubleshooting on the Light Microscopy Module (LMM) in the Fluids Integrated Rack on November 16. He tried replacing cables on the LMM. Nespoli worked on space plumbing throughout the day in the Waste and Hygiene Compartment (WHC). He removed and replaced valves and sensors in the WHC as part of regular preventative maintenance. Acaba unpacked food, batteries and computer equipment from the Cygnus.

Expedition 53, Week Ending November 26, 2017

Cubesats were launched from the Kibō module on November 20, to investigate antibiotic resistance, astrophysics and space weather. Nespoli and Ryazanskiy used the Sarcolab-3 experiment to look at leg muscles. Nespoli was the subject in the chair.

Nespoli tested a sensor attached to his forehead to measure his temperature continuously over 36 hours. The Circadian Rhythms data is allowing researchers to understand how astronauts cope with living in unnatural day-night cycles.

Further cubesats were launched from the Kibō module on November 21. TechEdSat demonstrated spacecraft and payload deorbit techniques. OSIRIS-3U cubesat recorded the Earth's ionosphere with the Arecibo Observatory in Puerto Rico.

Bresnik, Nespoli and Acaba entered the Bigelow Expandable Activity Module (BEAM) to start converting the module into a cargo platform by replacing old BEAM hardware with new electronics and stowage equipment.

One final cubesat group launched this week was the EcAMSat, which researches how the E. coli pathogen reacts to antibiotics in space.

---

[4] Gene Cernan was a *Gemini 9A, Apollo 10* and *Apollo 17* astronaut who was the second-last person to walk on the Moon for the first time. (He left the Lunar Module before Harrison Schmitt), but he was the last person to walk on the moon. (He entered the Lunar Module after Schmitt.) He died on January 16, 2017, aged 82.

Expedition 53, Week Ending December 3, 2017

Acaba and Misurkin tested a pair of SPHERES in the Kibō laboratory in preparation for a high school programming competition in January. Progress 67 fired its engines and raised the station's orbit on November 20.

While Nespoli, Ryazansky and Bresnik prepared to depart the station, the next crew to launch was in Russia for the traditional ceremonies before heading to the launch site in Kazakhstan. Anton Shkaplerov, Scott Tingle and Norishige Kanai were in Star City talking to journalists before heading to Moscow to tour Red Square and lay flowers at the Kremlin Wall where famed cosmonauts are interred. Meanwhile, the crew on the station packed the Cygnus craft with trash before it was to leave the station.

Expedition 53, Week Ending December 10, 2017

Ground controllers unberthed the Cygnus from the Unity module on December 5 and conducted communications tests as part of the Commercial Crew Program. They used the Canadarm2 robotic arm to detach the Cygnus spacecraft from the Earth-facing side of the station's Unity module. It was then placed near the Harmony module to gather data overnight. Vande Hei and Acaba controlled the Canadarm2 to release Cygnus back into Earth orbit at 8:10 a.m. EST on December 6. Later Cygnus released 14 cubesats from an external NanoRacks deployer.

Bigelow Aerospace and NASA signed a contract extension to allow BEAM to stay attached to the station for another 3 years, with a potential to stay an extra year. BEAM would not yet transition to a cargo hold, as engineers study its ability to resist radiation, space debris and microbes.

Expedition 53, Week Ending December 17, 2017

On December 13, Expedition 53 Commander Randy Bresnik of NASA handed over station command to Roscosmos' Alexander Misurkin in an official Change of Command ceremony. At 9:02 p.m. EST, the hatch closed between the Soyuz spacecraft and the ISS in preparation for undocking. Bresnik, Nespoli and Ryazanskiy undocked 12:14 a.m. EST to begin their trip home. They landed safely at 3:37 a.m. EST southeast of Dzhezkazgan in Kazakhstan.

Postscript

At the time of writing, Vita was Paulo Nespoli's last spaceflight.

Paolo Nespoli retired as an active astronaut at ESA in 2019 after his Vita mission to the ISS in 2017 as part of Expedition 52/53.

Fig. 9.10. Paolo Nespoli at Cork Institute of Technology, Cork, Ireland, July 2018. (CIT)[5]

He has received the following honors:

- NASA Spaceflight Medal, 2007
- Commendatore Ordine al Merito della Repubblica Italiana, 2007
- Cavaliere dell'Ordine della Stella della solidarietà Italiana, 2009

During his three space missions, Nespoli accumulated 313 days, 2 hours and 36 minutes in space.

---

[5] The author attended.

10

Horizons

Mission

ESA Mission Name:	Horizons
Astronaut:	Alexander Gerst
Mission Duration:	196 days, 17 hours, 50 minutes
Mission Sponsors:	ESA
ISS Milestones:	ISS 55S, 92nd crewed mission to the ISS

Launch

Launch Date/Time:	June 6, 2018, 11:12 UTC
Launch Site:	Pad 1, Baikonur Cosmodrome, Kazakhstan
Launch Vehicle:	Soyuz MS
Launch Mission:	Soyuz MS-09
Launch Vehicle Crew:	Sergei Valerievich Prokopyev (RKA), CDR
	Alexander Gerst (ESA), Flight Engineer
	Serena Maria Auñón-Chancellor (NASA) Flight Engineer

Docking

Soyuz MS-09

Docking Date/Time:	June 8, 2018, 13:01 UTC
Undocking Date/Time:	December 20, 2018, 01:40 UTC
Docking Port:	Rassvet nadir

Landing

Landing Date/Time:	December 20, 2018, 05:02 UTC
Landing Site:	Near Dzhezkazgan, Kazakhstan
Landing Vehicle:	Soyuz MS
Landing Mission:	Soyuz MS-09
Landing Vehicle Crew:	Sergei Valerievich Prokopyev (RKA), CDR
	Alexander Gerst (ESA), Flight Engineer
	Serena Maria Auñón-Chancellor (NASA) Flight Engineer

© Springer Nature Switzerland AG 2020
J. O'Sullivan, *European Missions to the International Space Station*,
Springer Praxis Books, https://doi.org/10.1007/978-3-030-30326-6_10

ISS Expeditions

ISS Expedition:	Expedition 56
ISS Crew:	Andrew Jay Feustel (NASA), ISS-CDR
	Oleg Germanovich Artemyev (RKA), ISS-Flight Engineer 1
	Richard Robert Arnold (NASA), ISS-Flight Engineer 2
	Sergei Valerievich Prokopyev (RKA), ISS-Flight Engineer 3
	Alexander Gerst (ESA), ISS-Flight Engineer 4
	Serena Maria Auñón-Chancellor (NASA), ISS-Flight Engineer 5
ISS Expedition:	Expedition 57
ISS Crew:	Alexander Gerst (ESA), ISS-CDR
	Sergei Valerievich Prokopyev (RKA), ISS-Flight Engineer 1
	Serena Maria Auñón-Chancellor (NASA), ISS-Flight Engineer 2
	Oleg Dmitriyevich Kononenko (RKA), ISS-Flight Engineer 3
	David Saint-Jacques (CSA), ISS-Flight Engineer 4
	Anne Charlotte McClain (NASA), ISS-Flight Engineer 5

The ISS Story So Far

Soyuz MS-06 delivered Aleksandr Misurkin, Mark Vande Hai and Joseph Acaba to the station in September 2017.

Soyuz MS-07 carried Russian Anton Shkaplerov, American Scott Tingle and Japanese Norishige Kanai to become part of Expedition 54 in December 2017. This was followed by Soyuz MS-08, which delivered Oleg Artemyev, Andrew Feustel and Richard Arnold to the station. They became part of Expedition 55 in March 2018.

Alexander Gerst's Previous Mission

See the Blue Dot mission.

The Horizons Mission

Mission Patches

The logo was designed to be timeless and shows a face gazing into the horizon above a blue band that symbolizes the atmosphere but also the Blue Dot mission. The new mission went beyond Blue Dot and extended into infinity symbolized by the white arc. To the right of the mission name is a stylized International Space

Station, which also represents a four-masted ship. The design incorporates the "golden section," where something is divided into minor and major sections and the proportion of minor to major is the same as major to whole. In this case, the golden section was used to measure the Earth/space ratio as well as the under orbit/over orbit ratio and many more measurements. It was designed by Christian K. Pfestorf and Michael Hartmann of the Rhein Main International Institute for Advanced Design Strategies.

The circular shape of the Soyuz MS-09 crew patch symbolizes the shape of the globe, inside of which the Belukha Mountain is visible – the highest peak of the Altai, representing the crew's call sign. A trio of white swans flying over Earth towards the International Space Station symbolize the three crew members. The six stars shown against the black background of space are for the members making up the successive ISS expedition crews. The golden silhouette of the orbital outpost is shown at the top of the emblem. Below are the numbers of the main expeditions of the same hue. The Roscosmos logo is placed in the middle of the inscription "Soyuz MS-09." The crew members' names are along the emblem's inner border, with the crew commander's name underneath the flying spaceship Soyuz. The outer border is made up of the astronauts' national flags. It was designed by Luc van den Abeelen.

Fig. 10.1. Horizons mission patch (ESA)

Fig. 10.2. Soyuz MS-09 mission patch (www.spacefacts.de)

Fig. 10.3. Expedition 56 mission patch (NASA)

Fig. 10.4. Expedition 57 mission patch (NASA)

The Expedition 56 patch portrays a dove carrying an olive branch in its beak. The patch includes images of the Soyuz launch vehicle for the crew and the ISS orbital laboratory. The Expedition 56 astronauts' names are displayed on the dove's wings and along the limb of Earth at the base of the patch. The dove's tail is firmly planted on Earth's limb in order to represent the strong link between our planet and the humans who are sent into the cosmos to carry us beyond the bounds of our only home. The patch theme is best expressed with the Sanskrit word "Shanti," meaning peace or tranquility in one of the oldest languages in history. The patch illustrates our hope for peace and love in the world, and the innate human desire to spread our wings and explore into the future, building on the wisdom of the past, for the betterment of humankind.

The Expedition 57 patch is a tribute to human exploration. It depicts an explorer's ship leaving for the unknown as our early ancestors did and is shaped like an arrow, heading out to new cosmic horizons. It highlights the purpose of the International Space Station as a world class science laboratory for the benefit of mankind and international cooperation, as well as humanity's flagship in space, preparing us for the amazing voyages ahead. The Expedition 57 patch is dedicated to all those thousands of humans who make this journey possible through the contribution of their passion, hard work, and courage to one of the most fascinating projects in human history.

Fig. 10.5. The Soyuz MS-09 crew with Alexander Gerst on right (www.spacefacts.de)

Fig. 10.6. The Expedition 56 crew with Alexander Gerst second from right (NASA)

Fig. 10.7. The Expedition 57 crew with Alexander Gerst in center. (NASA)

Timeline

Expedition 56, Week Ending June 10, 2018

Alexander Gerst wrote in his ESA blog on June 5:

> *With every new day, I am looking more and more forward to returning to the International Space Station – and doing somersaults while brushing my teeth again…*
>
> *I am also looking forward to all the scientific research up there. The Space Station is an extremely productive laboratory: on average, the crew supervise around 300 experiments for every six-month mission, from medicine and physics to engineering and life sciences. There is no other place than in space that these experiments can be conducted. The insights that emerge from these unique experiments are immensely important for the preparation of future missions into space, but also for everyday life here on Earth. This combination makes research in orbit so valuable to each of us – and to the generations that follow us.*

These benefits for people on Earth are clear for some experiments: for example, we cannot examine the properties of liquid metal alloys on Earth without gravity disrupting the process. As soon as we pour liquid alloys into a container on Earth, the material interacts with the container on an atomic scale. To avoid this container-interference you could let the alloy float in mid-air, but this is only possible on the International Space Station. The Electromagnetic Levitator is a special smelting furnace I installed in the Columbus module during the Blue Dot mission, and it allows us to observe precisely how alloys behave. In this way, we can determine their physical properties to calculate and simulate future alloys that will be stable and lighter and can be made and used on Earth. We do not perform a full research investigation in space, but instead we close the small – but crucial – gaps in our knowledge that have been blocking us scientifically on Earth.

These types of experiments have already made possible the development of new, fuel-efficient and quieter aircraft turbine-blades made of titanium aluminide. In the Horizons mission, we are now investigating transparent materials that we can look into and observe the dynamics of their crystal formation.

We are also experimenting with tiny semiconductors that may help develop the next generation of computer chips on Earth. We measure how foams can be designed so that stable, resource-saving materials can emerge from them – on Earth foams disintegrate quickly due to gravity (like the foam in a glass of beer). We are also exploring the quantum mechanics of extremely cold accumulations of atoms called "Bose-Einstein condensates." These condensates are not stable long enough to observe on ground: but findings from the Space Station could be instrumental in helping us develop new, revolutionary computers.

In medical research, we astronauts are often our own research subjects: studies have shown similarities between the ailments we suffer from in weightlessness to the difficulties chronically ill people live with on Earth. Our immune system is weaker and weightlessness causes astronauts to lose muscle and bone mass; even our sense of balance needs to adjust to the new environment, much like people recovering from a stroke.

Astronauts are healthy test subjects, however, allowing researchers to investigate what exactly happens to our bodies on the International Space Station at a cellular level and what we can do to stop the negative effects. For example, the Brain-DTI experiment compares MRI images of our brain structures before and after the mission. We take saliva, urine and blood samples to track changes in the immune system. In the Microgravity Science Glovebox, we inject new anti-cancer drugs into tumour cells that float weightlessly in the samples, forming spherical structures – which is a far better simulation of cells in the human body than lying flat in a Petri dish.

It will also be crucial for astronauts to cope with the medical challenges of weightlessness regarding the limited resources that a spaceship could carry along on longer missions in space, such as to the Moon and Mars. We already recycle many resources on the International Space Station, including almost all of our drinking water; nevertheless, we still need regular deliveries from Earth.

We do not yet have a life support system that is good enough to fly to Mars: closing the recycling loop completely is very hard to achieve. That is why I am particularly pleased with a new photo-bioreactor that will be installed and tested during the Horizons mission. It uses algae to treat the station's air supply, and is a good example of progress being made in this direction. On Earth, recycling is becoming more and more important as we need to use the limited resources we have more efficiently.

With some other experiments it is hard to see at first glance what concrete benefits will come from them, but fundamental research is of utmost importance, because ground-breaking insights regularly come from it – insights nobody could foresee in advance. Among the important research tasks of my mission are the small, seemingly simple experiments that I will carry out for students and schoolchildren. During the Blue Dot mission, I experimented with soap bubbles, paper airplanes and gyroscopes in weightlessness and sent films of these experiments to ground. I would like to continue building on these educational experiments in the Horizons mission. That might even be the most important thing I can bring back from my mission in space: inspiring the next generation of explorers.

I would be happy if as many boys and girls think to themselves: "What he does up there, I can do, too – and if I try my best, I can probably do even more." Children grow mentally when shown possibilities – if such a spark ignites in the astronauts, scientists, and engineers of tomorrow, I have successfully accomplished my mission.

Soyuz MS-09 launched from the Baikonur, Kazakhstan at 7:12 a.m. EDT on June 6. At the time of launch, the station was flying about 250 miles above south-central Egypt. On board were Serena Auñón-Chancellor of NASA, Alexander Gerst of ESA and Sergey Prokopyev of Roscosmos. The Soyuz MS-09 spacecraft docked to the Rassvet module at 9:01 a.m. EDT on June 8. Hatches were opened at 11:17 a.m. EDT and the new crew members were welcomed by Expedition 56 Commander Drew Feustel and flight engineers Ricky Arnold and Oleg Artemyev.

Expedition 56, Week Ending June 17, 2018

Alexander Gerst spoke to the European Astronaut Centre in Cologne, Germany, at 17:25 CEST on June 12. He answered questions about climate change, World Cup football and how it feels to return to space. Arnold investigated and issue with a semiconductor crystal growth experiment in the Microgravity Science Glovebox. Feustel moved a pair of incubator units to a new location to make room for new experiments being delivered on the next SpaceX Dragon cargo mission. Finally, they readied the Quest airlock and their spacesuits for Thursday's EVA.

Gerst worked on experiments studying the effects of living and working in space on June 13. He assumed a "face-up" position for the Grip study researching the nervous system. The results may improve the design of future spacecraft and help patients on Earth with neurological diseases. He later installed sample equipment in the Electromagnetic Levitator furnace.

Feustel and Arnold of NASA started their spacewalk at 8:06 a.m. EDT on June 14, when they switched their spacesuits to battery power. They installed new high-definition cameras ahead of the commercial crew SpaceX Crew Dragon and Boeing Starliner commercial crew spacecraft, expected to begin operations in 2020. They performed some quick tasks such as replacing a camera on the starboard truss, closing an aperture door on an external environmental imaging experiment on the Kibō laboratory and relocating a grapple bar. The EVA ended at 2:55 p.m. EDT and lasted 6 hours, 49 minutes. This was the 211th spacewalk in support of assembly and maintenance of the station.

Expedition 56, Week Ending June 24, 2018

On June 19, Auñón-Chancellor collected blood and urine samples before working with Gerst on the Myotone study observing how microgravity affects the biochemical properties of muscles. Gerst took part in the Everywear experiment, attempting to make astronauts more efficient with their time. Auñón-Chancellor, Gerst and Prokopyev reviewed the location of safety equipment and trained for emergency communication in the station's Russian segment. Arnold cleaned a burner in the Advanced Combustion Microgravity Experiment. Feustel worked in the Quest airlock, clearing up the previous spacewalk and scrubbing the water loops in the spacesuits.

The NanoRacks-Remove Debris satellite was deployed on June 20, from the Kibō module. This satellite tracked space debris with a 3D camera and captured another nanosatellite in a deployable net as a proof of concept. Arnold continued to troubleshoot the Microgravity Science Glovebox semiconductor crystal growth experiment. Gerst measured acoustic levels with dosimeters. Arnold, Feustel and Artemyev later conducted an emergency evacuation drill.

Gerst participated in a survey on time perception of space crew on June 21. Then he measured his pulmonary function while using the exercise bike. Arnold, Feustel and Auñón-Chancellor had eye exams with an ultrasound device to study microgravity's effects on eyesight. Auñón-Chancellor replaced a filter and a valve in the station's bathroom in the Tranquility module. She tested newly installed Wi-Fi antennas and helped Feustel stow equipment after the earlier successful deployment of a nanosatellite.

Expedition 56, Week Ending July 1, 2018

Feustel began the new week stowing biological samples for the Multi-Omics study. He also worked on the atomization experiment investigating how fuel is sprayed in the combustion chambers of jet and rocket engines. Auñón-Chancellor installed a new cubesat deployment system in the Kibō laboratory containing nine nanosatellites.

On June 26, Feustel and Arnold practices with the Canadarm2 in advance of the arrival of the SpaceX Dragon capsule. Feustel cleaned the Rodent Research-7 experiment and Arnold verified that the Veggie facility was ready for the Veg-03 study. Auñón-Chancellor studied the microstructures of solidifying cement in microgravity. This is to aid the design of future lightweight space structures, for example, on Mars. Gerst worked on spacesuit equipment in the Quest airlock for future spacewalk operations. He purged nitrogen from the suit's oxygen lines and set up an overnight oxygen leak check.

The following day, Auñón-Chancellor prepared the Microgravity Science Glovebox for the Angiex Cancer Therapy experiment, due to arrive on the next Dragon. This cargo ship was launched on a SpaceX Falcon 9 rocket at 5:42 a.m. EDT on June 29, from Cape Canaveral Air Force Station, Florida. It carried more than 5900 pounds of research investigations and equipment, cargo and supplies. Included in the cargo were several experiments to be installed in the ESA Columbus laboratory. These included the two payloads for the International Commercial Experiments service (ICE Cubes). Hydra-2 studies the microbes' reactions to microgravity and explores their potential as bio-miners to produce methane. Hydra-3 merges art and science by using a person's heart rate to change a piece of kaleidoscopic artwork. Also, on board was the intelligent mobile crew assistant, CIMON. This is an autonomous free-flying assistant.

Three other experiments were receiving upgrades and new equipment. They were MAGVECTOR, which investigates how Earth's magnetic field interacts with an electrical conductor, the Soft Matter Dynamics instrument works with foams, emulsions and granular matter and finally the Compacted Granulars experiment (CompGran), which monitors the behavior of colliding grains until they reach complete arrest. The Dragon's unpressurized trunk also carried a new Latching End Effector for the Canadarm2.

Expedition 56, Week Ending July 8, 2018

Arnold and Feustel grappled the Dragon spacecraft with the Canadarm2 at 6:54 a.m. EDT on July 2. Ground controllers then berthed the Dragon to the Harmony module. Two more experiments arrived on the Dragon: Micro-12 and ECOsystem Spaceborne Thermal Radiometer Experiment on Space Station (ECOSTRESS). The former studies how electrons are transferred across the cell membrane of bacteria. The latter examines how plants grow with varied water availability by studying the Earth's surface.

The crew spent the fourth of July holiday on light duty.

Overnight on July 6, ground controllers commanded the Canadarm2 robotic arm to extract a new Earth-observing experiment from the rear of the Dragon. The new ECOSTRESS equipment was then remotely installed on the outside of the Kibō laboratory. Auñón-Chancellor and Gerst continued with the Myotones muscle study by taking blood samples.

Expedition 56, Week Ending July 15, 2018

The Progress 70 cargo craft launched at 5:51 p.m. EDT on July 9 from Baikonur, Kazakhstan, while the station was about 250 miles over southwest Uzbekistan. The station was over the Tasman Sea between Australia and New Zealand, two orbits later, when the cargo ship docked at 9:31 p.m. EDT to the Pirs Docking Compartment. Launch to docking was only 3 hours and 40 minutes, a new record for supply ships arriving at the station. With its arrival, there were now six spacecraft attached to the ISS: Progress 69, the Soyuz MS-08, Soyuz MS-09, SpaceX Dragon and Northrop Grumman Cygnus.

The thrusters on the Cygnus cargo ship fired for 50 s on July 10 at 4:25 p.m. EDT to reboost the station. This test proved that the Cygnus could be used, as well as the Progress, to raise the station's altitude, if required. The station was raised by about 295 ft.

Auñón-Chancellor and Gerst each worked on cancer studies. Gerst used a microscope to look at proteins that could be used for cancer treatment and radiation protection.

July 11 was a day of unloading. Auñón-Chancellor unloaded science equipment and station hardware from inside the SpaceX Dragon, while Artemyev and Prokopyev continued unloading the nearly 3 tons of crew supplies and station hardware delivered aboard the Progress 70 cargo craft.

The next day, Auñón-Chancellor continued cancer research as she examined endothelial cells as part of the AngieX Cancer Therapy study. This is looking for a better treatment to target tumor cells. Then she helped Feustel with the Micro-11 fertility study, researching whether successful reproduction is possible in space.

Gerst and Auñón-Chancellor commanded the Canadarm2 to release the Cygnus cargo spacecraft at 8:37 a.m. EDT on July 15, while the complex was 253 miles above the southeastern border of Colombia.

Expedition 56, Week Ending July 22, 2018

On July 17, the crew retrieved, recharged and redeployed Aerosol Sampling Experiment in Nodes 1 and 3. These samplers collect particles in the air through thermophoresis, a process in which different particle types exhibit different responses to the force of a temperature gradient. The crew installed six space algae culture bags in the Veggie facility.

Alexander Gerst wrote in his ESA blog on July 18:

We, the crew of Soyuz MS 09, arrived safely on the ISS, started our research – and it feels fantastic to be back up here!

The launch with the Soyuz rocket was bombastic, but relatively relaxed nonetheless. What sounds like a contradiction has a very simple reason: because this time, I knew what to expect.

The atmosphere on board was fantastic, thanks to my crew mates Sergey and Serena. This showed, for example, as we sat in the narrow cockpit, waiting for pressure tests to be conducted ahead of our flight. Earlier, each crew member had been given the chance to request a few pieces of music to be played over the radio during the pressure tests. Serena's list included Michael Jackson's "Billy Jean," and when it was played, we all started dancing with our hands for fun. We thought we were unobserved, but at that very moment, Mission Control turned on our cockpit camera! I'm sure that caused a few spontaneous laughs in the control centre, not to mention all the viewers watching the live stream. The show was not planned, but it does illustrate an important point in our profession: one should always retain a sense of humour. There are enough serious moments.

Things also went almost too smoothly during our flight to the space station. Unlike in training, there was not a single moment during the real flight when Sergey and I had to test our limits as pilots and co-pilots. What is absolutely desirable, of course, appeared unusually quiet to us. Unlike the first mission, I had completed the full training programme to be able to control the Soyuz myself. And that was a great feeling: knowing that I really have this under control. I could fly a spaceship with my own hands if necessary, and not just in the simulator.

Even in automatic mode you have to control the spaceship with countless inputs on the console. However, compared to the simulator, in the real spaceship you typically always check two or three times whether you are actually entering the right commands. Pressing just two wrong buttons would be enough to blow the spaceship in three parts or depressurize it immediately, without an "are you really sure?" This sounds dangerous, and it is, but in some situations, such as an emergency re-entry or fire, we have to do just that in a flash to save the life of the crew.

By the way, our training under a steady bombardment of malfunctions was not completely useless. During our approach to the ISS, shortly before a thrust manoeuvre, an alarm signal indicated that one of our engines had failed. However, thanks to the safe design of the Soyuz capsule, this was no problem: we simply switched over to the spare engine for the next thrust. And, in the end, it actually turned out that our thrusters were all fine, only a sensor had triggered incorrectly.

It is in these moments when our training really pays off. You have to be constantly alert, constantly acting, even if the spaceship often flies automatically. The most experienced cosmonauts rightly say you have to "fly ahead" of the spaceship in your mind, ready to manually take over at any moment.

The two days of living in the tiny Soyuz were long, but contrary to my expectations I slept relatively well. Once I slept for eight hours and in the next "night" I slept almost eleven hours – what a luxury compared to the previous months. And luckily, like on my first mission, I didn't get sick, even though the capsule is in a spin during most of the approach to. After these two rather calm days I was completely relaxed when we arrived on ISS, having had the time to eat, go to the toilet and get used to life in space.

When the hatch opened to the station, it felt like coming home to me. I immediately got along as well with floating as a fish does with water. Interestingly, my feet and hands have even unconsciously remembered where handrails and foot loops are, and usually, when I flew around a corner, they grasped the correct handrail on their own. But there was one time it didn't work out so well. My body thought it knew exactly where a handle had to be and reached out to grab it – only to find someone had moved it since my last mission…

I still don't know exactly why, but my body appears to have been created specifically for such strange environments. Just as I always felt very comfortable under water as a child, in free fall while skydiving, or in the cold of Antarctica, for some strange reason it feels completely natural for me to float around here in space.

During my first mission, the biggest challenge of my first few weeks was re-familiarising myself with the many complex computer systems and pieces of equipment on ISS. But now even that feels as if I never left, which is lucky as we had to start preparing difficult scientific experiments straight away. Serena and I also had a complicated onboard assignment to support two NASA colleagues during their spacewalk, right in our first week on board.

For that assignment, we manually had to manoeuvre the massive ISS robotic arm, and also we had to help our colleagues Drew and Ricky into their spacesuits, performing the final check before they would leave the relative safety of ISS in order to work outside in open space. When putting on helmets, shoes and gloves, every little mistake can be fatal, since the environment in open space is not exactly life-friendly. Like a shepherd in a stormy night, I only relaxed again once my two colleagues were safely back inside.

I am very much looking forward to the months ahead in this fantastic laboratory, with hundreds of unique scientific experiments that will make life on Earth better. Thanks to my experience I can now help my new colleagues settle in up here. But the learning phase is not over for me either. Because when you work inside the most complex machine of humankind, there is always something new to learn.

And that's a good thing.

Feustel and Auñón-Chancellor worked on the Micro-11 fertility study again on July 20. Feustel worked with the Amyloid experiment which is aiming to develop advanced treatments for Alzheimer's disease and diabetes. He collected amyloid fibril samples from the Cell Biology Experiment Facility and stowed them in a science freezer for spectroscopy and microscopic analysis back on Earth. Gerst and Arnold took air quality samples and swabbed the stations interior surfaces as a part of microbe investigations. Gerst put samples in the freezer for later analysis and Arnold processed microbial DNA using the Biomolecule Sequencer. This is with a view to establishing how microbes survive in microgravity.

Gerst was beamed live to the Jazz Open Festival on Stuttgart's Schlossplatz, at 21:50 local time. "Good evening, Kraftwerk, good evening, Stuttgart!" he greeted the electronic band Kraftwerk and 7500 attendees at the festival. He said "I am one of only six people in space, on the outpost of humanity, the International Space Station ISS, 400 km above sea level. The ISS is a manmade machine, the most complex and valuable machine humankind has ever built. Here in the European Columbus laboratory, the successor to the Spacelab, the European Space Agency ESA is researching things that will improve daily life on Earth. More than 100 different nations work together peacefully here and achieve things that a single nation could never achieve. We are developing technologies on board the ISS to grow beyond our current horizons and prepare to take further steps into space, to the Moon and Mars." Gerst then played a synthesiser on a tablet and accompanied the band in a rendition of their song, Spacelab.[1]

Expedition 56, Week Ending July 29, 2018

The Progress 69 fired its engines on July 26 to increase the station's orbital altitude. Feustel and Auñón-Chancellor continued a second week of research on fertility in space. Gerst examined the sedimentary properties of quartz and clay particles which could help design geological missions on planetary missions. Arnold set up the Aerosol Samplers in the Harmony and Tranquility modules while Gerst scanned Arnold's eyes as part of the Optical Coherence Tomography study.

Expedition 56, Week Ending August 5, 2018

While Auñón-Chancellor worked on the AngieX Cancer Therapy and examined endothelial cells on July 30, Arnold loaded the Dragon with hardware and science samples. Gerst worked in the Columbus module to help doctors understand how an astronaut's perception of time and distance is affected during and after a mission.

---

[1] Spacelab is a track on the 1978 Kraftwerk album, *Die Mensch-Maschine* (*The Man-Machine*).

On August 1, Auñón-Chancellor and Gerst explored the hypothesis that upward fluid shifts in the body caused by microgravity increases pressure on the brain, possibly pushing against the eyes. This may affect the shape of the eye and permanently affect vision. Artemyev and Prokopyev prepared for their upcoming spacewalk by reviewing the translation paths to their work sites on the outside of the station's Russian segment.

The Dragon was released by the Canadarm2 on August 2.

A new era of human spaceflight was announced on August 3 when NASA revealed the names of the astronaut who would fly to the station on the first commercial crew vehicles. "Today, our country's dreams of greater achievements in space are within our grasp," said NASA Administrator Jim Bridenstine. "This accomplished group of American astronauts, flying on new spacecraft developed by our commercial partners Boeing and SpaceX, will launch a new era of human spaceflight. Today's announcement advances our great American vision and strengthens the nation's leadership in space."

NASA announced that Robert Behnken and Douglas Hurley were to fly on a SpaceX Dragon 2 in April 2019 and that Christopher Ferguson, Eric Boe and Nicole Mann were scheduled to fly on a Boeing CST-100 Starliner sometime in mid-2019.[2]

Expedition 56, Week Ending August 12, 2018

Artemyev and Prokopyev started the week preparing for their spacewalk. They installed batteries that will power their spacesuit, ensured their suits were sized properly and conducted leak checks.

Auñón-Chancellor shut down and stowed the AngieX Cancer Therapy experiment on August 7. Feustel and Arnold readied a pair of SPHERES satellites for a student competition. Middle school students in the United States competed to write the best algorithms to operate the SPHERES simulating a mission on Saturn's moon Enceladus. Gerst used a sextant with star maps as an emergency form of navigation in space.

Arnold began troubleshooting the Cold Atom Lab (CAL). This cools atoms to about one ten-billionth of a degree above absolute zero. He inspected the fiber cables of the equipment and he installed it in the Columbus laboratory. Feustel, Gerst and Arnold prepared the Kibō module for the upcoming arrival of the JAXA H-II Transfer Vehicle (HTV) in September. The HTV was scheduled to carry new lithium ion batteries for installation on the station's Port-4 truss power system.

---

[2] At the time of publication (Spring 2020) both Dragon 2 and CST-100 are scheduled to carry crew in mid 2020. Eric Boe was replaced by Michael Finke when he was deemed unable to fly for medical reasons.

Expedition 56, Week Ending August 19, 2018

Artemyev and Prokopyev loaded the Progress 69 in preparation for its departure then continued preparing for their EVA in the Pirs airlock. Feustel worked on a protein crystallization experiment while Auñón-Chancellor researched the radio spectrum utilized in space with a view to improving satellite communications. Arnold tested U. S. spacesuit lights and replaced fan filters. Gerst was also maintaining the spacesuits by inspecting the batteries. Later he tested the station's fire extinguishers and breathing apparatus.

On August 15, at 12:17 p.m. EDT, Artemyev and Prokopyev opened the hatch of the Pirs docking compartment and started their Russian segment spacewalk. They manually launched four small technology satellites and installed an experiment called Icarus onto the Russian segment of the space station. The EVA ended at 08:03 p.m. EDT.

The following is translated from Alexander's original German ESA blog text by the Horizons team:

We had a visit from Earth! The SpaceX Dragon CRS-15 resupply vehicle arrived Monday 1 July bringing provisions, new scientific experiments and a whole lot more work for the crew on board the International Space Station (ISS).

Thanks to this exciting arrival, the past few weeks have been even busier than usual. Ricky and Drew captured the capsule with the robotic arm, which is always a tricky manoeuvre and requires full concentration. Then we began the unloading.

Some materials and equipment had to be cooled immediately. And because the Dragon was also supposed to bring a whole series of completed experiments back to Earth, we had to hurry to get them done in time. As a result, we have spent more than 104 hours a week on research – just short of the ISS team record!

In the Columbus module I was responsible for commissioning the MagVector experiment, after a conversion that allows researchers to investigate how magnetic fields deflect cosmic radiation. I also installed a new camera in the Fluid Science Laboratory – a facility used to investigate the dynamics of foams and granules in weightlessness.

In the U. S. module, we've been cultivating algae and Arabidopsis plants, carrying out research into cancer cells and even mixing "space concrete." Here in weightlessness, where no disturbing convection fluxes occur in the material, we're able to produce particularly pure cement and concrete. The benefits of this are twofold. On the one hand, it helps us better assess the properties of these very important building materials and improve their production on Earth. On the other, we gain experience that could be very useful for us when it comes to constructing a future Moon base.

On top of all this research, we also had time for a very special exercise. It's one I had been looking forward to for a long time: Serena and I were to attempt to navigate by the stars from the ISS Cupola using a sextant.

As a huge fan of exploration, the idea that we would be using a sextant was extremely cool for me. Functioning like a protractor, it's a tool that allows you to determine your position based on the orientation of celestial bodies. This is not only possible on the ground but also up in space.

Navigating by sextant aboard the ISS may seem antiquated at first glance. But optical angle measurement was vital in helping the Apollo 13 *astronauts return to Earth after an explosion on board their spacecraft in 1970. And NASA and ESA want to equip us with this almost 300-year-old invention for future missions to the Moon or Mars for that very reason – in case of emergency.*

But, before they do so, it needs to prove its worth. So, from the Cupola, Serena and I tried to measure the angles between certain stars.

We quickly realized that the exercise planners, who know almost every star in the night sky inside out, were more concerned about technical questions – how the light comes in or where to attach the foot straps in order to assume a stable observation position. What they hadn't considered, however, was that the most important thing when using a sextant on board the ISS is having an exact star map for the planned observation time.

You can see so many stars through the windows of the ISS (especially on a moonless night) that you can hardly find your way around. The constellations no longer stand out, and the section of the "sky" is relatively small, making it rare to be able to see the entire constellation. Moreover, our Space Station rotates around itself at four degrees per minute, so you can only see each star for a few minutes at a time before it either disappears behind the Earth's horizon or is covered by part of the ISS structure. We managed quite well, but we still have some work to do to determine the best way forward.

This is often the case in space travel; in practice, completely different problems become more important than those that have been carefully considered beforehand. To close this gap, you need astronauts.

The sextant exercise is also a good example of how we as astronauts deal with the risks of our missions. We rely on many different safety grids, and develop a plan B, C, D and E for as many scenarios as possible.

For example, recently we had a power failure on board the ISS, during which one of eight solar array power channels collapsed early in the morning following the unfavourable impact of a cosmic elementary particle. A second channel had been deactivated earlier due to the low sun angle. And, because the current load was then transferred to the other circuits, a third channel also collapsed before the system was able to recover.

Fortunately, no vital systems – such as the cooling system of the station – were affected. They are always fed by several independent solar array wings (SAW). But the onboard alarm woke us pretty swiftly from our sleeping cabins. We couldn't coordinate with the ground control team because we were in a dead spot. So first we had to tackle the problem on our own. In pajamas.

It's at this point that we are reminded: the ISS is a highly complex system that is floating alone in the hostile cosmos. Even a power failure like this could rapidly result in serious consequences if there were not so many redundancies in place to ensure safety.

You can imagine our Space Station as a mobile that hangs over a baby's crib, with many colorful figures on it, in perfect balance. If you cut part of the mobile off, it will affect the rest and put it in danger of falling off balance. The art is to build the mobile in such a way that error chains do not spread so far, and that the largest part of the structure always remains stable.

In this respect, the ISS is very cleverly designed. It can re-balance itself within seconds by cutting off power supply to non-vital areas. But to get all the important components – such as life support, cooling, navigation and communication systems – back up and running, astronauts must intervene.

As astronauts, we train for a long time to keep calm in such critical situations and not lose sight of the system as a whole. It's important to adapt to the special, exposed environment up here not just physically but also mentally. From parachute jumping to deep diving underwater or a month-long research stay in extreme polar cold, our bodies are remarkably well-equipped to handle different environments and situations. It's the mind that matters the most.

What really fascinates me about these kinds of environments is not the danger, the isolation or the adrenaline. On the contrary, I don't like any of those three things very much. What I do like is the possibility of conquering, controlling and mastering such a hostile environment and using it for our scientific benefit. And then ending up feeling at home in a place that's so far from it.

That's what drives me on such expeditions – and what makes up for the risk or suffering up here in space.

Expedition 56, Week Ending August 26, 2018

Gerst exercised on August 20 wearing a t-shirt as part of the SpaceTex-2 study. According to the German Aerospace Centre (DLR), it is designed to "provide comfort, efficient thermal control and sweat evaporation during a workout in microgravity". Feustel set up the DLR Earth hyperspectral imagery equipment on the exterior of the Kibō laboratory. It monitors urban and rural areas, the health of vegetation and water areas as well as the environmental effects of natural and manmade disasters.

On August 22, Feustel worked on the BioServe Protein Crystalography study using a microscope. This is aiming to decrease the time needed to develop medicines.

The Progress 69 cargo craft undocked from the aft port of the Zvezda Service Module at 9:16 p.m. CDT, then moved to a safe distance from the orbital laboratory for the customary week of engineering tests by Russian flight controllers before it was commanded to deorbit.

Auñón-Chancellor carried out the twice-yearly checkup of the treadmill in the Tranquility module on August 24. She checked the treadmill's belt tension, greased axles and replaced parts. Feustel set up a pair of the SPHERES satellites for the SmoothNav experiment. This tests the autonomous operation of the SPHERES and tests algorithms for them to communicate with each other. Arnold reinstalled the Plant Habitat after some maintenance work on the Kibō EXPRESS rack. He also installed new software in the METEOR experiment laptop and installed a camera in the Destiny laboratory's Window Observational Research Facility (WORF).

Expedition 56, Week Ending September 2, 2018

Five of the six crewmembers participated in the regularly scheduled eye exams and vision checks on August 27. They tested visual acuity with an eye chart. Auñón-Chancellor and Feustel used an ultrasound device to examine their optical nerves and retinas. Arnold and Gerst checked eye pressure with a tonometer. Arnold extracted RNA to help researchers decipher the changes in gene expression that take place in microgravity. Feustel photographed protein crystal samples with a microscope to help doctors develop more effective disease-treating drugs on Earth. Gerst exercised again wearing the DLR SpaceTex-2 T-shirt. He also participated in the GRIP study investigating if cognitive ability is affected by manipulating objects in space.

On August 28, Feustel and Auñón-Chancellor practiced robotics techniques for the arrival of the H-II Transfer Vehicle (HTV-7), scheduled for September 14.

At about 7 p.m. EDT on August 30, flight controllers in Houston and Moscow detected a pressure leak in the complex. They decided not to wake the crew. Once the crew members were awake, they began to investigate the leak and tried to find it. It was traced to the Russian segment. It was quickly isolated to a hole about 2 millimeters in diameter in the orbital compartment of the Soyuz MS-09 spacecraft docked at the Rassvet module of the Russian segment. This was the Soyuz that carried Prokopyev, Gerst and Auñón-Chancellor to the station in June. More worryingly, it was the Soyuz that was to take them back to Earth in December. The crew slowed the rate of the leak through the temporary application of Kapton tape.

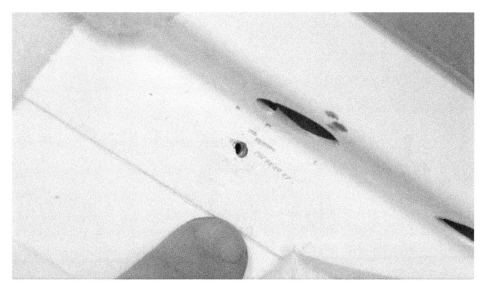

Fig. 10.8. Hole inside the orbital compartment of Soyuz MS-09. Marks show where the drill slipped. (NASA)

Flight controllers in Houston and Moscow worked together with the crew to affect a repair. Prokopyev of Roscosmos used epoxy on a gauze wipe to plug the hole and flight controllers in Moscow performed a partial increase of the station's atmosphere using the Progress 70s oxygen supply. Flight controllers in Moscow and Houston continued to monitor station's cabin pressure. Meanwhile, Roscosmos has convened a commission to conduct further analysis of the possible cause of the leak.

The crew resumed a regular schedule of work on August 31 as the station's cabin pressure was holding steady. Feustel prepared tools to be used in a pair of spacewalks the following month.[3]

Expedition 56, Week Ending September 9, 2018

Feustel installed a new camera controller and data storage unit on the Electromagnetic Levitator (EML) in the Columbus laboratory on September 5. Arnold worked in the Kibō laboratory replacing valves inside the EXPRESS Rack-5. The rack contains experiments operated by astronauts on the station and

---

[3] Theories of sabotage by a homesick crew member or impact by micrometeorites were dismissed. It was determined that the Soyuz capsule was damaged by workers on the ground and a botched repair was attempted and covered up. Drill marks near the hole were made by a type of drill not on the station. Sealant used to fill the hole dried out in orbit and hence the leak occurred.

scientists on the ground. Gerst worked with German health studies, testing another t-shirt which includes a sensor for cardio-pulmonary diagnosis during exercise. Gerst, Feustel and Auñón-Chancellor reviewed the plan for the arrival of the HTV-7 resupply ship.

The Zvezda service module fired its engines for 13 seconds to boost the station's orbit in advance of a pair of Soyuz crew ships departing and arriving in the coming month.

Expedition 56, Week Ending September 16, 2018

Feustel and Arnold examined the station's resident mice on September 10 for the Rodent Research-7 (RR-7) experiment. They checked the breathing and mass of the rodents before placing them back in their habitat and restocking their food. RR-7 is observing how microgravity impacts gut microbes and how it may affect astronaut health. Gerst finished an experiment before finalizing spacesuit work in the U. S. Quest airlock. He stowed science equipment in the morning that analyzed the exhaled air of astronauts to detect signs of airway inflammation. Later he completed the battery charging of the U. S. spacesuits then began regenerating metal oxide canisters in advance of a pair of spacewalks at the end of the month.

Expedition 56, Week Ending September 23, 2018

The Zvezda service module was used again on September 20 to raise the station's orbit. Arnold and Auñón-Chancellor conducted ultrasound scans inside the Columbus laboratory. Arnold processed blood and urine samples using the Human Research Facility's centrifuge. Feustel installed computer network equipment in an EXPRESS rack in the Unity module, while Gerst moved smoke detectors in the Tranquility module then performed routine maintenance on laptops in the Destiny laboratory.

JAXA's H-IIB rocket launched at 1:52 p.m. EDT on September 22 after a number of postponements due to weather. When it launched from the Tanegashima Space Center in southern Japan the space station was 254 miles over the southwestern Pacific, west of Chile. A little more than 15 minutes after launch, the unpiloted H-II Transfer Vehicle-7 (HTV-7) cargo spacecraft successfully separated from the rocket and began its 4.5 hours rendezvous with the station.

Expedition 56, Week Ending September 30, 2018

As the crew prepared for the arrival of the HTV cargo ship, Auñón-Chancellor sequenced DNA from microbial samples collected in the station on September 24. Feustel worked on an experiment to atomize liquids with a view to improving fuel efficiency, both on Earth and in space. Feustel, Artemyev and Arnold also began preparations for their return to Earth, planned for October. Artemyev and Feustel practiced their descent on a computer simulator, and Arnold stowed crew items in the Soyuz.

Feustel and Auñón-Chancellor opened the Bigelow Expandable Activity Module (BEAM) for maintenance and stowage on September 25. They reinforced structural elements in the inflatable module to extend its operational lifespan and stowed equipment.

Feustel and Auñón-Chancellor grappled the JAXA Kounotori H-II Transfer Vehicle (HTV-7) at 7:34 a.m. EDT on September 27 and completed capture at 7:36 a.m. At the time of capture, the space station and cargo spacecraft were flying 250 miles above the northern Pacific Ocean. Robotic ground controllers then berthed the HTV-7 on the Earth-facing side of the Harmony module. The seventh Japanese cargo ship was loaded with more than 5 tons of supplies, water, spare parts and experiments. The cargo included six new lithium-ion batteries, a new sample holder for the Electrostatic Levitation Furnace (JAXA-ELF), a protein crystal growth experiment at low temperatures (JAXA LT PCG), a Life Sciences Glovebox, and EXPRESS racks.

Expedition 56/57, Week Ending October 7, 2018

On October 1, Frank de Winne, head of the ESA European Astronaut Centre in Cologne, Germany, was interviewed on the subject of commanding the ISS.[4] This was in anticipation of Alexander Gerst taking command of the station on October 3 with Luca Parmitano set to take command in the coming year:

> *'Of course, you are always extremely happy when something like that happens, first of all personally, because it is certainly recognition of your capabilities, but also at an organisational level. In this case, it was recognition of the progress ESA had made over many years of investment in human spaceflight. We were the first of the international partners, outside of the USA and Russia, to take on the position, and it really spoke to the strength we had developed within the European Astronaut core and the European human spaceflight community.'*

> *Decisions around which roles each astronaut will fulfil during a mission are made in consensus by a Multilateral Crew Operations Panel (MCOP). This panel comprises representatives from NASA, Roscosmos, ESA, JAXA and Canada, and roles are often assigned more than two years in advance. While overall command of the station lies with flight directors on the ground unless there is an emergency on board. Frank says a key role of the International Space Station commander is to foster team spirit. This involves ensuring crew members are able to perform the tasks required of them during their time in space.*

> *'It is about making sure that every crew member is happy – that everyone understands their role and contributes to the best of their abilities so the team can deliver optimal performance for the program,' he explains.*

---

[4] Belgian astronaut Frank De Winne commanded the station during Expedition 21 in 2009.

In many ways, Frank believes the position of Space Station commander is similar to other leadership roles on Earth, and he is 100 percent confident both Alexander and Luca will do a stellar job.

'*I have watched Alexander evolve over the past six years – from his first space flight to the time he was assigned as a commander. I have seen how he has grown into his role.*

'*Having Luca take up the position during his mission as well says a lot about Europe's position as a trusted partner. People can rely on us, and they do rely on us. Not only in terms of the hardware that we provide to the space station and now the service module for Orion, but also in the area of crew operations.*

'*I think that is a very good result of the investment ESA member states have made.' This sentiment is echoed by Alexander who says: 'I am humbled and honoured to command the International Space Station. This international sign of trust reflects Europe's reliability in space, and was made possible by the fantastic work of my European colleagues on their previous missions.*

'*For me, becoming the Space Station commander for the first time is a learning experience and something that I have had to work hard for. It might be scary at the beginning, but then you grow into that position. And at the end, you realize that it was much easier than you thought.'*

All leadership positions have their challenges. For Frank, many of these were logistical as the first commander of a six-person Space Station crew.

'*Up until the very last moment there was a risk that the number of crew members would be reduced. I had many meetings with the International Space Station programme manager and his deputy to discuss how we could help with the logistics situation on board and in the transport vehicle to ensure everyone could fly.*

'*Another highlight came when we were still on the ground. At that time, the Space Station crew was still involved in the tasking panel, and some of my crew members did not have the high-level tasks I would have wanted them to have. I worked a lot with the tasking panel to make sure there was a good balance between tasks such as robotics and EVA (extra vehicular activity) for all members of my crew.'*

In addition to these challenges, Frank says living with six people in a confined space means always being aware of any stress or unhappiness building up. You must also be ready to take command of the station and direct the crew at any time in an emergency event.

Because everyone has their own leadership style, Frank says the conversations he has had with Alexander have been about transferring knowledge rather than advice.

'Everybody fulfills the role with their own character and experience, but what I could do was tell Alex as much as I could about my experiences and what I remembered. I was able to let him know what worked and what did not work so well for me, and I am sure once he returns from his mission, he will be able to do the same.'

Alexander will take over the role of International Space Station commander from NASA astronaut Drew Feustel on 3 October. This marks the start of expedition 57 and second part of Alexander's Horizons mission.

Alexander will remain in the role of Space Station commander until he returns to Earth in December 2018.

Alexander Gerst also wrote himself about his recent experiences on the station in his ESA blog on October 1:

Every astronaut knows that we are in a potentially life-threatening situation on the space station – surrounded by the cosmic vacuum – but at the same time we feel relatively safe up here, compared to our hostile environment.

There are countless precautions on the ISS to minimize sources of danger. We trained for thousands of hours to deal with emergencies properly. And in everyday life, we are often so absorbed in the work on our scientific experiments – because you forget for a while that we are not floating through a research laboratory on Earth but in the cosmos.

In some moments, however, the exposure up here is abruptly brought to our minds. For example, four weeks ago, when the ground control informed us early in the morning after getting up by radio that we had a leak.

In the station, the air pressure dropped very slowly over the course of days, then ever faster – fortunately, even so slowly, that the air reserves would have been enough for four more days. But in such situations, the long emergency training really pays off: we immediately switched over the head and quickly tracked down the leak with our flight controllers in Moscow, Houston and Oberpfaffenhofen. In the orbital module of our Soyuz capsule docked to the ISS, which is blasted to the ground and burned up on the return flight, there was a hole about 3 mm in diameter in the outer wall – fortunately in a location that was accessible to us.

We then sealed the leak – first with a finger, then with adhesive tape, later permanently with a graft of parts of a gauze bandage and epoxy resin. We will be able to continue our work and fly safely back to Earth at the end of our mission. At the moment it is still being investigated how the hole in the launch preparation process of the spaceship was created exactly.

But for us the important thing is: everything under control.

Our preparation has proven itself. Often, risk researchers say that, under time pressure, we humans are not good at thinking rationally about complex situations of danger associated with many imponderables. We do not overlook her. And our brain can only process a limited number of impressions at the same time.

Everything that you have not trained before is a separate burden. Like driving a car, for example: getting in, strapping on, shoulder-glance, clutching, navigating, listening to the passenger; and maybe the radio is on at the same time. All these things clog up the capacity to think about them separately. This can probably be understood by every driver from his time as a novice driver.

In stress situations, our perception capacity continues to drop. I know that from situations of some of my colleagues skydiving. In extreme cases, it can happen that only one single impression can be processed, which has already led to some accidents: you only see the ground, which comes closer and closer, without you having to think about what to do next, you become incapacitated. Of course, we absolutely have to avoid such a 'tunnel view' for astronauts here on the space station.

That's why there are so many procedures that can be used in emergencies. And that's why we train so long that many moves become routine and do not burden our heads. We can then resort to familiar patterns. This frees up the perception capacities that we need when it matters.

In reality, of course, often everything is quite different. Many situations are so complicated and tricky that no protocol of action could guide us completely there. Exactly in this case, it is important that we keep the overall view and in addition to use our gut feeling that we have developed from experience.

Such moments are the hardest for a commander, because he has to decide in no time and under stress when to deviate from the procedures – and when to return to them. Particularly tricky is the so-called 'Flash of Brilliance,' a supposedly brilliant, spontaneous idea to carry out an emergency plan other than planned. However, there is a great risk that a great idea at first sight may later lead to a dangerous trap.

The art of leadership in an emergency situation is therefore for the commander to take the haste out of a situation, to exude calm, to make the right decision with information and intuition – or to act immediately when the situation requires it.

How important this is, I have experienced before, when I was in my hometown Künzelsau with the volunteer fire department. It gives you an intuition of where hidden dangers might lurk. Today, I also benefit from this on the space station: If in doubt, you can use a respirator extra, or double-check that there are no misunderstandings within the crew.

Some risks persist despite all preparations. But one must also admit: to the discovery of the unknown is one of them. The U. S. entrepreneur and philanthropist John Shedd once summed up the beginning of the 20th century very well, I think, when he said, 'For ships, the safest way is to stay in the harbour. But ships are not made for that.' And T. S. Eliot's quote stems that only those who run the risk of going too far are able to figure out how far they can go.

If we are not prepared to make some imponderables, we will not get anywhere in life, and even in space. Of course, even if one cannot really weigh personal and global risks against each other, the biggest dangers for humanity are ultimately to do nothing at all. If we do not explore the universe around us, we will never be able to protect ourselves from the risks we face, for example, from meteorite impacts or space radiation. And without the satellite view from the outside, we would never have discovered such global, 'internal' threats to our lives as climate change.

For myself, I have spent a lot of time with the risks involved with my mission here in space and decided: I accept it.

Not everything up here is cool and fun. A life in space, even on a large space station, is full of hardships. But I like to work, because I know that it fills my life with meaning. It helps me to clarify the big questions for me, which probably everyone will face at some point: What is my place in life, in this world? What do I give back to society from what I've heard?

I had a lot of luck, a nice, safe childhood in Germany, a great family, many choices. I never had to go hungry. And I think it also creates a responsibility to do something to help other people, in our own and in other countries that are not doing so well, have a better life.

As an astronaut, I can help better heal diseases, better understand our Earth, protect it, develop better materials, inspire the next generation of researchers. That makes it easy to be here in space. As I have already felt on research missions in the Antarctic and on volcanoes, so I'm sure of something here in space: Here I found my place in the world. At least until the next opportunity to explore the world beyond my own horizon.

An ESA astronaut took command of the International Space Station for the second time since 2009 on October 3. NASA's Drew Feustel handed over command to Alexander Gerst in the traditional Change of Command Ceremony. Belgian astronaut Frank De Winne commanded the station during Expedition 21 in 2009.

Expedition 56 commander Drew Feustel and flight engineers Ricky Arnold and Oleg Artemyev undocked from the International Space Station at 3:57 a.m. EDT on October 4. Artemyev commanded the Soyuz for its journey home. They landed at 7:44 a.m. EDT southeast of the town of Dzhezkazgan in Kazakhstan.

On October 8, Gerst explained his goals for Expedition 57 during a live event at the International Astronautical Congress held in Bremen, Germany. He said "My personal goal as a space station commander is to have all of our experiments done, to keep this fantastic research platform in a good state and to come back home as friends."

Expedition 57, Week Ending October 14, 2018

On October 10, Soyuz MS-10 launched from the Baikonur Cosmodrome in Kazakhstan 4:40 a.m. EDT. 121 seconds after lift-off, the craft went into contingency abort due to a sensor failure and executed a ballistic descent back to Earth. Aleksei Ovchinin and Nicholas Hague were recovered in good health. They landed 402 km from the launch site and 20 km east of Jezkazgan, Kazakhstan The Soyuz MS-10 flight abort was the first Russian crewed booster accident at high altitude in 43 years, since Soyuz 18A failed to make it to orbit in April 1975.[5]

NASA Administrator Jim Bridenstine, who was at Baikonur to watch the launch, made a statement on twitter: "NASA astronaut Nick Hague and Russian cosmonaut Alexey Ovchinin are in good condition following today's aborted launch. I'm grateful that everyone is safe. A thorough investigation into the cause of the incident will be conducted."

Expedition 57, Week Ending October 21, 2018

Auñón-Chancellor tested battery life as part of the Zero G Battery Test experiment. New station commander, Gerst checked the carbon dioxide and water management in a life support rack while Prokopyev likewise maintained the life support systems in the Russian segment. He also ran on the treadmill in the Zvezda service module.

After the failed launch of Soyuz MS-10, the International Space Station partners held meetings on October 15 and reshuffled the tasks for the three astronauts on the station. They postponed two planned spacewalks as Nick Hague was scheduled to join Gerst for both spacewalks.

Gerst logged 300 cumulative days in space across his two missions, Blue Dot and Horizons, on October 16. At this point, for European astronauts, only Thomas Reiter with 350 days and Paolo Nespoli with 313 days had spent more time in space. Gerst was destined to pass both these records during Horizons, ending the mission with 362 days in space.

Auñón-Chancellor relocated samples collected from biology experiments into a Kibō science freezer on October 17. Then she researched how to grow protein crystals real-time on the space station. Gerst carried out maintenance on the station toilets and worked on equipment inside the Destiny laboratory module before

---

[5] Vasili Lazarev and Oleg Makarov survived the descent. Makarov went on to fly in space twice more on-board Soyuz 27 and Soyuz T-3, both flights to the Salyut 6 space station.

updating a warning procedures book. Prokopyev explored how forces such as exercising or spacecraft dockings impact the station's structure. He then participated in a study observing interactions between a space crew and Mission Control in Moscow.

Expedition 57, Week Ending October 28, 2018

Auñón-Chancellor continued to process microbial samples gathered form the air and surfaces of the station on October 22. These samples would later go through genetic sequencing to establish how microgravity has affected the microbes. She also photographed protein crystals while Gerst replaced hardware inside the Combustion Integrated Rack in the Destiny laboratory. He photographed further quartz and clay sedimentation. Prokopyev set up the EarthKAM experiment in the Unity module to allow school pupils to remotely operate a camera aimed at the Earth's surface.

Auñón-Chancellor replaced camera cables and a laptop in the Multi-Purpose Small Payload Rack (MSPR) in the Kibō laboratory on October 23. She tested breathing apparatus connected to station oxygen supply with Gerst. Then they stowed unwanted items in the HTV-7 resupply ship.

On October 25, Auñón-Chancellor prepared Veggie plant growth facility in the Columbus laboratory to grow kale and lettuce. Gerst photographed the station's port side solar arrays so that ground specialists find any damaged or worn areas. Prokopyev checked on power supply systems in the Zarya cargo module.

Gerst scrubbed the cooling loops on a pair of spacesuits inside the Quest airlock on October 26. Auñón-Chancellor and Prokopyev cleaned the Electrostatic Levitation Furnace (ELF) and the Combustion Integrated Rack (CIR). Prokopyev then replaced CIR fuel tanks.

Expedition 57, Week Ending November 4, 2018

Auñón-Chancellor took part in the NeuroMapping experiment on October 29. This examines the neuro-cognitive abilities before, during and after a spaceflight. In this session, she took test where she examines rotating objects, attempts to hit targets while moving and concentrates on two tasks at the same time. Gerst continued the GRIP study in the Columbus laboratory. He performed tasks while strapped into a seat and gripping a sensor device. Prokopyev set up the SPHERES satellites for a test in the Kibō laboratory.

On October 31, Auñón-Chancellor replaced cartridge holders in the Electrostatic Levitation Furnace (ELF). Gerst configured the Life Sciences Glovebox in the Kibō laboratory. Prokopyev replaced the fuel tanks in the Combustion Integrated Rack's (CIR) in the Destiny laboratory.

On November 1, revised crew assignments and cargo shipments were announced after the failure of Soyuz MS-10 to reach orbit on October 11. Roscosmos planned to launch the Progress 71 resupply mission on November 16 and announced that Soyuz MS-11 would launch on December 3 with Oleg Kononenko, David

Saint-Jacques of the Canadian Space Agency and NASA's astronaut Anne McClain on board.

Roscosmos completed an investigation into the loss of the Soyuz when one of four first stage rocket engines abnormally separated and hit the second stage rocket that led to the loss of stabilization. A state commission confirmed that a damaged separation sensor caused the launch abort. According to the launch sequence of the Soyuz rocket, around 2 minutes into the flight the booster should detach from its mounting. A sensor activates a valve that ejects gas to push away the top of the booster. Since the sensor did not work, one of the boosters was not pushed away but instead slipped down and damaged the rocket's second stage. The investigative commission stated that the "launch concluded with a failure of the launch vehicle due to abnormal separation of one of the side boosters, whose nose section hit the core stage in the area of the fuel tank, which led to its depressurization and the loss of stability of the launch vehicle."

The Roscosmos statement said: "The reason for the abnormal separation is the non-opening of the nozzle cap of the 'D' block oxidizer tank because of the deformation of the stem of the separation contact sensor (bending on 6° 45'), which was admitted when assembling the 'package' at the Baikonur Cosmodrome. The cause of the LV accident is of operational nature and extends to the backlog of the 'Soyuz' type LV 'package.'"

NASA and ESA released a 3 minutes clip featuring time-lapse footage of Earth and video of scientific experiments performed on board, shot by Gerst. It was the first 8K video shot aboard the International Space Station. Andrea Conigli, of ESA's video production and photography team, said the introduction of 8K camera technology is about more than just resolution. It complements the growing capability of astronauts to document and share their experiences with the world. "With so much video available, audiences expect an increasingly high level of quality. We're constantly coming up with new ideas as to how we can engage, educate and spark curiosity among the next generation of innovators and explorers," Andrea said.[6]

Expedition 57, Week Ending November 11, 2018

Gerst and Auñón-Chancellor packed the HTV-7 cargo ship on November 6. Prolopyev examined the weightless environment and sunlight affect plasma-dust crystals. He photographed the interior of the Russian segment then updated the station's inventory system.

---

[6] Previous Hi-Def videos have included productions like Samantha Cristoforetti's interactive panoramic tour of the station, Thomas Pesquet's short film 'New Eyes,' and ESA's first ever 4K video comprised of imagery captured by Alexander during his Blue Dot mission in 2014. Luca Parmitano was trained to use a RED 8K camera on his Beyond mission.

Gerst, with the support of Auñón-Chancellor, used the Canadarm2 robotic arm to release the HTV-7 cargo spacecraft at 11:51 a.m. EST on November 7. At the time of release, the space station was flying 254 miles over the northern Pacific Ocean. As part of an engineering test, a new, small, re-entry capsule was deployed by HTV-7. This 2-foot high and 2.7 foot wide conical capsule was designed by JAXA and assembled by the station crew to test the ability to return small payloads from the station for expedited delivery to researchers.

Expedition 57, Week Ending November 18, 2018

Gerst and Serena practiced the robotic capture of the Cygnus spacecraft. Gerst examined the new free-flying robotic assistant, CIMON, or Crew Interactive MObile CompanioN. CIMON was developed and built by Airbus in Friedrichshafen and Bremen, Germany, on behalf of the German Space Agency DLR and uses artificial intelligence software by IBM Watson. Ludwig Maximilians University Clinic in Munich (LMU) is in charge of the project's scientific aspects. It weighs around 5 kg and has a display screen at its center. It will support the crew by displaying and explaining information needed to carry out scientific experiments and maintenance. It can take photographs and record video to record tasks or find items. It is the space station equivalent of Alexa or Siri.

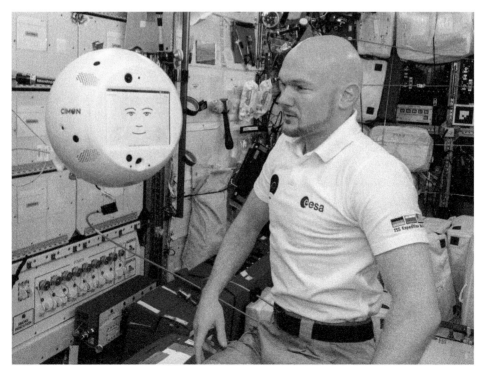

Fig. 10.9. Alexander Gerst with CIMON, November 2018 (DLR)

Auñón-Chancellor worked on the Life Sciences Glovebox then moved on to orbital plumbing tasks. Prokopyev primarily worked in the Russian segment on life support maintenance and science experiments.

Gerst and Prokopyev trained for the upcoming arrival of Progress 71, practicing manually docking the vehicle. Gerst also practiced the grappling of the Cygnus spacecraft which cannot dock automatically.

Carrying almost 3 tons of food, fuel and supplies, the Progress 71 cargo spacecraft launched at 1:14 p.m. EST on November 16 from Baikonur.

The following day, Northrop Grumman's Antares rocket, carrying the Cygnus NG-10 cargo spacecraft, lifted off at 4:01 a.m. EST from the Mid-Atlantic Regional Spaceport, on Wallops Island, Virginia. This was the first Cygnus redesignated after Northrop Grumman purchased Orbital ATK in June 2018.

Progress 71 cargo ship docked at 2:28 p.m. EST on November 18 to the aft port of the Zvezda Service Module.

Expedition 57, Week Ending November 25, 2018

Auñón-Chancellor used the Canadarm2 to grapple the Cygnus NG-10 spacecraft, "SS John Young"[7] before ground controllers commanded the station's arm to rotate and install the Cygnus on the bottom of the station's Unity module. It brought nearly 7400 pounds of research and supplies to the station.

Gerst spent most of November 19 unpacking the Cygnus supply spacecraft. The following day he took photos for NASA's Binary Colloidal Alloy Test. On November 21 he installed new modules for the NanoRacks equipment and answered the "At home in space" questionnaire that is researching how future space stations could be more accommodating to astronauts. On November 23 he collected samples from around the station for analysis. As the outpost is a closed system, any dust and microbes quickly multiply, and the environment needs to be checked regularly.

The International Space Station turned 20 years old on November 20, 2018. The first element, the Zarya module, was launched on November 20, 1998. See Appendix "Alexander Gerst FAQ" in this book for 20 frequently asked questions about Alexander Gerst's time on board the ISS, published by ESA to celebrate the anniversary.

Auñón-Chancellor and Gerst opened Cygnus's hatch, installed new science freezers and transferred cargo, while Prokopyev opened the Progress 71 hatch and began unloading new equipment. Gerst also photographed more clay and quartz

---

[7] John Young is the only spacefarer to have commanded four different types of spacecraft: Gemini, Apollo Command and Service Module, Apollo Lunar Module and space shuttle. He flew on the first Gemini and the first space shuttle. He orbited the Moon on *Apollo 10* and he walked on the Moon during *Apollo 16*. He retired from NASA in 2004, aged 74, but continued to attend Astronaut Office Monday morning meetings for years.

samples. Auñón-Chancellor measured light intensity in the Columbus laboratory as part of the replacement of fluorescent lights with LEDs and the effect on crew wellness.

Expedition 57, Week Ending December 2, 2018

Gerst scanned the eyes of Auñón-Chancellor on November 26 with an ultrasound device. He then observed protein crystals as part of a study on Parkinson's disease. Auñón-Chancellor set up hardware for a semiconductor crystal experiment. Prokopyev transferred fluids to the Progress 70 cargo craft.

On November 28 Gerst wrote in his ESA blog:

Due to the incident on the Soyuz MS-10 flight, which our colleagues Alexey Ovchinin and Nick Hague should have brought to the ISS in October, Serena, Sergey and I will now only be here in space for a few weeks longer than planned: probably until the 3rd of December, until the next crew joins us.

Of course, we had looked forward to the two, we had even prepared their sleeping cabins and set them on the day of the launch orange juice in the fridge.

It is a pity that our friends are not with us. But we are not shocked. It has always been clear to us from the beginning that such anomalies can occur. We train for scenarios that are far worse. And one has to say: It's impressive to see how well the rescue system works and what the spaceship has done. The Soyuz is so ingeniously designed that even in such a critical case it could save the crew. The two have arrived with only the slightest bruises on the ground. The more than 1,000 search and rescue experts stationed below the thousands of kilometres long start trajectory at every astronautic rocket launch from Baikonur have surely saved them: this shows how international space teamwork works when it counts.

Even the three of us up here at the station see our new situation above all as a positive challenge. It depends on the attitude! Presumably we will not be able to perform all the experiments planned for the ISS Expedition-57. We also had to postpone the spacewalks first.

But we'll be fine, and we're well taken care of. As much as I have always been tempted to travel out into hostile spaces and defy them (whether in Antarctica, at the edge of volcanoes, or in space), the three of us now say we have difficult conditions, but we have made the most of it – now for good. And in the end we will be able to look back with pride on what we have achieved.

The most important thing is that we do not hang ourselves down now, that no one is lethargic or overwhelmed. That's why we pay more attention to one another and do more together: for example, we now always eat in the evening, often at noon, in threes. Every Saturday we watch a movie together on our projector screen in the "Knoten 2," everyone can choose a movie in turn. Recently, for example, I showed a two-parter about the "Endurance" expedition led by Ernest Shackleton in the Antarctic: It's good to see other explorers master much more difficult situations in the unknown.

During the day we always run cameras in the research modules, so that the ground control can take more part in our work and we remain attentive. And in return, I asked our control centre to stream a live image of the control room to one of our laptops. With the experiments we get along surprisingly well, we hardly lag behind. And the mood is quieter, but actually better than before. In a sense, the special situation also helps us to motivate ourselves: now it depends on each of us fully, no one can retire.

Sometimes we celebrate it correctly. For example, each of us has some so-called bonus food containers up here, for which she or he was allowed to personally select the content. Normally, every crew member eats this stuff for the most part alone. But now we always share everything, and each one of us sometimes floats by the others to bring them an extra piece of chocolate or a cappuccino.

I think I'm so lucky to have a crew with such a great attitude here! Of course, as a commander, I make sure that everything goes as well as possible, both the experiments and our living together in space everyday life. For that I have to spend more time than before. I talk to my flight director on a regular basis about the tasks, work overtime in the evenings, for example to clean up, to organize things or for repairs.

But I like to do it. As the leader of a polar expedition, as a commander, I am responsible for always keeping the overview – and for doing so, I also have to pocket one for the team.

We still have a lot of work to do. Now several space transports arrived, which we had to catch, empty and reload. And the end of our mission, when the next crew arrives, is sure to be really dynamic.

Countless tricks to get along well in everyday life here on the space station are nowhere written down. They are handed down from team to team: how to best pack the trash, tie up your sleeping bag, go to the bathroom, wash, how to put on a space suit in weightlessness, what screw it is, how to hoop 20 hovering tools at the same time One must treat with special care how to photograph a sunrise, which corners of the Earth are particularly photogenic.

One can imagine the ISS as a big house, with many technical peculiarities and systems, which of course sometimes gets in the way. Anyone who has lived in it for a long time knows exactly how to use it and how to solve many tricky situations. But who moves in new, has no idea of these peculiarities – and cannot prepare for it in advance.

There are a thousand small ideas that some crew on the ISS has at some point solved a problem. But writing all this down would be impossible. It's a space-life culture that has evolved over the years.

Normally, a handover lasts for at least two months to explain all the gimmicks of the next crew, but we will only have two weeks to spare, while the new ones will have to get used to the challenges of space at the same time.

Serena, Sergey and I are currently the only carriers of the ISS culture in space. What we forget to tell our followers in this extremely short time will probably be lost forever as knowledge. That is a big responsibility that weighs on us!

My two colleagues Sergey and Serena, who are on the ISS for the first time, do not know this situation yet; and it's great to see how they are already accepting this challenge and preparing to help the newcomer.

The time here is still like the calm before the storm. But even that may be a good finding from the Soyuz MS-10 incident: He has made everyone aware again that astronautical space missions are not routine events. We work at the limits of the technically feasible in order to carry them further outward. Our missions are riddled with challenges that we need to focus on for the benefit of humanity. Even if they are not easy, or better, as John F. Kennedy formulated it at the beginning of the Moon program, precisely because they are not easy, but difficult.

On November 29, Prokopyev checked the Orlan spacesuits that he and Kononenko planned to wear during their upcoming spacewalk. Gerst and Auñón-Chancellor helped in performing leak-checks. Gerst and Auñón-Chancellor then scanned their head and foot muscles with an ultrasound device in the Columbus laboratory.

Expedition 57, Week Ending December 9, 2018

The Soyuz MS-11 launched from the Baikonur, Kazakhstan at 6:31 a.m. EST Monday, December 3. NASA astronaut Anne McClain,[8] David Saint-Jacques of the Canadian Space Agency, and Oleg Konenenko of Roscosmos were onboard

---

[8] In a divorce and custody battle, McClain's wife Summer Worden accused McClain of identity theft, i.e. accessing her bank account from a computer on board the ISS. This was the first allegation of a crime in space.

and ready to join Expedition 57. The Soyuz spacecraft docked to the ISS's Poisk zenith port at 12:33 p.m. EST. The hatches between the Soyuz spacecraft and the station opened at 2:37 p.m. EST and the trio were welcomed by Expedition 57 Commander Alexander Gerst, Serena Auñón-Chancellor and Sergey Prokopyev. This was the first spaceflight for both McClain and Saint-Jacques and the fourth trip to the space station for Kononenko.

On December 5, SpaceX Dragon CRS-16 launched from Cape Canaveral SLC-40 at 18:16 UTC carrying over 5600 pounds of science, supplies and hardware for the crew.

Gerst later worked on U. S. spacesuit maintenance cleaning their cooling loops. Serena worked on a cement study inside the orbital laboratory that could inform the construction of future lunar or Martian habitats. McClain, Kononenko and Saint-Jacques were familiarizing themselves with the station. Kononenko also joined Prokopyev to prepare a pair of Russian Orlan spacesuits for an upcoming spacewalk to inspect the Soyuz MS-09, which would return Prokopyev, Gerst and Serena back to Earth. It has sprung a leak.

Gerst and Auñón-Chancellor captured the Dragon spacecraft at 7:21 a.m. EST on December 8, while the station was over Papua New Guinea; then the ground controllers took over and installed the spacecraft on the bottom of the station's Harmony module.

Fig. 10.10. Alexander Gerst in the Copula while capturing the SpaceX Dragon capsule, December 2018 (NASA)

Expedition 57, Week Ending December 16, 2018

Kononenko and Prokopyev started their spacewalk at 10:59 a.m. EST December 11 when they opened the hatch of the Pirs docking compartment. During the 7 hours 45 minutes spacewalk, they examined the looked at the exterior of the Russian Soyuz MS-09 spacecraft and took photographs. They cut away the original external thermal blanket until they found the hole and confirmed that the sealant applied had reached the exterior of the hole. They also retrieved science experiments from the exterior of the Rassvet module before ending the EVA. It was the 213th spacewalk in support of ISS assembly, maintenance and upgrades, the fourth for Kononenko, and the second for Prokopyev.

Kononenko and Prokopyev maintained their Russian Orlan spacesuits on 12 December, after their spacewalk the previous day. Gerst and Serena Auñón-Chancellor drew blood samples and processed them in the Human Research Facility's centrifuge. Gerst packed the Soyuz spacecraft ready for the return journey to Earth with Auñón-Chancellor and Prokopyev. Gerst continued unpacking the SpaceX Dragon cargo craft and packing the Soyuz MS-09 on December 13. Auñón-Chancellor processed research samples in the NanoRacks Plate Reader that enables pharmaceutical and biotechnology science in space. She also stowed biological samples in a science freezer for a cellular adaptation study.

Expedition 57, Week Ending December 23, 2018

Alexander Gerst transferred command of the ISS to cosmonaut Oleg Kononenko on December 18. The following day, at 5:30 p.m. EST, the hatch closed between the Soyuz spacecraft and the station. Gerst,

Auñón-Chancellor and Soyuz commander Sergey Prokopyev undocked from the ISS at 8:40 p.m. EST to begin their trip home. They landed safely at 12:02 a.m. EST in Kazakhstan.

Postscript

At the time of writing, Horizons was Alexander Gerst's last spaceflight. For more on Gerst, see the Blue Dot mission.

11

Beyond

ESA Mission Beyond

Astronaut:	Luca Parmitano
Mission Duration:	Ongoing at the time of writing (September 2019)
Mission Sponsors:	ESA
ISS Milestones:	ISS 59S, 95th crewed mission to the ISS[1]

Launch

Landing Date/Time:	July 20, 2019, 16:28 UTC
Launch Site:	Pad 1, Baikonur Cosmodrome, Kazakhstan
Launch Vehicle:	Soyuz MS
Launch Mission:	Soyuz MS-13
Launch Vehicle Crew:	Aleksandr Aleksandrovic Skvortsov (RKA), CDR
	Luca Salvo Parmitano (ESA), Flight Engineer
	Andrew Richard Morgan (NASA) Flight Engineer

Docking

Soyuz MS-13

Docking Date/Time:	July 20, 2019, 22:47 UTC
Docking Port:	Zvezda aft

ISS Expeditions

ISS Expedition 60

ISS Crew:	Aleksei Nikolaevich Ovchinin (RKA), ISS-CDR
	Tyler Nicklaus Hague (NASA), ISS-Flight Engineer 1
	Christina Hammock Koch (NASA), ISS-Flight Engineer 2
	Aleksandr Aleksandrovich Skvortsov (RKA), ISS-Flight Engineer 3
	Luca Salvo Parmitano (ESA), ISS-Flight Engineer 4
	Andrew Richard Morgan (NASA), ISS-Flight Engineer 5

---

[1] 95th successful mission. 96th attempt after Soyuz MS-10 was aborted on 11 October 2018. Aleksei Ovchinin and Nick Hague landed safely after an emergency separation and ballistic trajectory landing.

© Springer Nature Switzerland AG 2020
J. O'Sullivan, *European Missions to the International Space Station*,
Springer Praxis Books, https://doi.org/10.1007/978-3-030-30326-6_11

ISS Expedition 61	
ISS Crew:	Luca Salvo Parmitano (ESA), ISS-CDR
	Aleksandr Aleksandrovich Skvortsov (RKA), ISS-Flight Engineer 1
	Andrew Richard Morgan (NASA), ISS-Flight Engineer 2
	Christina Hammock Koch (NASA), ISS-Flight Engineer 3
	Oleg Ivanovich Skripochka (RKA), ISS-Flight Engineer 4
	Jessica Ulrika Meir (NASA), ISS-Flight Engineer 5

The ISS Story So Far

Soyuz MS-10 launched on October 11, 2018, carrying Aleksey Ovchinin and Nick Hague. They were intended to join the crew of Expedition 57, however minutes after launch a booster failed to separate correctly, recontacted the core and was aborted. The crew capsule landed successfully after undergoing a ballistic trajectory. They had reached 93 km altitude and landed 19 minutes and 41 seconds after launch. The Soyuz-FG rocket was cleared to fly and Ovchinin and Hague launched successfully on Soyuz MS-12 on March 14, 2019, accompanied by Christina Koch of NASA. All three joined Expedition 59.

Between Soyuz MS-10 and Soyuz MS-12, Soyuz MS-11 carried Oleg Kononenko (RKA), David Saint-Jacques (CSA) and Anne McClain (NASA) to the station to join Expedition 57.

Luca Parmitano

See the Volare mission.

The Beyond Mission

Mission Patches

In choosing the name Beyond for the mission, Luca was inspired by his fellow ESA astronauts. From the nearness to Earth of Thomas Pesquet's Proxima mission to the broadening scope of Alexander Gerst's current Horizons mission, Luca saw a path that will push humankind even farther, for the benefit of all. "What we do in orbit is not just for the astronauts or for the International Space Station programme, it is for everybody," Luca explains. "It is for Earth, it is for humankind, and it is the only path for us to learn what we need in terms of science and technology in order to go beyond."

The mission logo illustrates this trajectory. An astronaut looks out into space through a helmet visor, with Earth and the International Space Station reflected in the visor. In the distance, the Moon is poised for humankind's return, with the Orion spacecraft and exploratory rovers. Beyond is Mars, the Red Planet, currently being studied by satellites such as ExoMars and Mars Express and, one day, by humans.

Fig. 11.1. Beyond mission patch (ESA)

Fig. 11.2. Soyuz MS-13 mission patch (www.spacefacts.de)

Fig. 11.3. Expedition 60 mission patch (NASA)

Fig. 11.4. Expedition 61 mission patch (www.spacefacts.de)

The Soyuz MS-13 patch was designed by Luc van den Abeelen of spacepatches. nl. He describes the design: "The circular Soyuz MS-13 crew patch focuses on the Soyuz launcher, depicted sitting on the Baikonur launch pad, while the service tower is opening as the spaceship is preparing to be launched into orbit. With the craft enclosed in the rocket nosecone, the launcher scene is surrounded by silhouetted stages of the Soyuz's lifecycle during its mission to the International Space Station. To the left is the ship approaching the orbital facility, which itself is depicted at the top of the design, while the missions end is illustrated by the parachute-assisted landing of the descent module. Three gold stars to the left of the ISS represent the spaceship crew, consisting of Russian commander Skvortsov, NASA astronaut Morgan and ESA astronaut Parmitano. Their names are positioned on the left while the Roscosmos logo is shown on the right of the design, underneath the spacecraft's designation".

The Expedition 60 crew insignia of the Moon landing symbolizes one of the most extraordinary feats of humankind, an embodiment of ingenuity and desire for exploration. The patch of Expedition 60 commemorates the 50th anniversary of that landing; a constellation of three stars with the Moon superimposed forms the letter "L," the Latin symbol for 50. The Moon is depicted as a waxing crescent, as it was on July 20, 1969. The familiar silhouette of the International Space Station is visible, flying across the night sky. Stars, numerous and bright as seen from the space station, form the shape of an eagle in the same pose as on the iconic patch of the *Apollo 11* mission. The sunrise represents the fact that we are still in

Fig. 11.5. The Soyuz MS-13 crew with Luca Parmitano on right (www.spacefacts.de)

Fig. 11.6. The Expedition 60 crew with Luca Parmitano in center right (www.spacefacts.de)

Fig. 11.7. The Expedition 61 crew with Luca Parmitano third from left (www.spacefacts.de)

the early stages of humanity's exploration of space. The hexagonal shape of the patch represents the space station's cupola, with the six points of the hexagon symbolizing the six crew members of Expedition 60. The names and nationalities are not present, as on the original *Apollo 11* mission patch, to highlight that space missions – then, now, and in the future – are for Earth and all humankind.

The Expedition 61 patch is a simple, yet striking design with LXI in Roman numerals set in the blue Earth and a sun illuminating a symbolic ISS. The crew names are in an offset ring.

Timeline

Expedition 60, Week Ending July 21, 2019

Exactly 50 years after Neil Armstrong and Buzz Aldrin became the first humans to step on the Moon, Andrew Morgan, Luca Parmitano and Alexander Skvortsov launched at 12:28 p.m. EDT July 20 from the Baikonur Cosmodrome in Kazakhstan. The Soyuz spacecraft docked at the station at 6:48 p.m. EDT while both spacecraft were over southern Russia, northeast of the Black Sea. They were welcomed by Expedition 60 Commander Alexey Ovchinin, Nick Hague and Christina Koch when the hatches opened at 9:04 p.m. EDT.

Expedition 60, Week Ending July 28, 2019

With the Expedition 60 crew complete with six members, the crew prepared for an upcoming spacewalk. Parmitano, Hague and Koch serviced U. S. spacesuits on July 22. Ovchinin photographed the Zvezda service module docking port and checked radiation levels. Skvortsov took an inventory of the cargo that arrived on the Soyuz.

Station commander Ovchinin checked the air quality in the Russian side of the station on July 23. Skvortsov checked the hatch seal and recharged batteries in the new Soyuz MS-13 crew ship docked to the Zvezda service module. At the end of the day, the entire crew reviewed their roles and responsibilities in the event of an emergency. They reviewed procedures, safety gear and escape paths for unlikely emergency scenarios such as a fire or a pressure leak aboard the station.

Morgan filmed himself on July 24 with a 360° camera in the Harmony module after Parmitano had set up the camera to record the virtual reality experience. Morgan, Parmitano and Skvortsov familiarized themselves with safety procedures and the station's layout. Koch set up the Astrobee robot and calibrated and tested its free-flying motion. Later, she took RNA samples from a freezer and put them in the Biomolecule Sequencer. Hague photographed the window in the Destiny laboratory. This was to check for cracks, scratches and contamination. Ovchinin and Skvortsov configured the Soyuz MS-13 for it stay at the station.

The High Definition Earth-Viewing (HDEV) experiment failed to communicate and transmit, and ground engineers started to investigate.

Koch recorded Morgan and Parmitano while they worked in the Columbus laboratory on July 25. She then measured airflow throughout the U. S. segment. Hague replaced urine recycling tanks in the Tranquility module and stored algae samples in the science freezer. He checked the new 360° camera. Ovchinin and Skvortsov worked on cardiac activity studies in the Russian segment, then while Ovchinin maintained communications equipment, Skvortsov continued unloading the Soyuz updating the station's inventory system.

The SpaceX Dragon CRS-18 mission launched at 6:01 p.m. EDT, July 25 from Cape Canaveral Air Force Station in Florida, carrying more than 5000 pounds of research, hardware and supplies. Hague grappled the Dragon at 9:11 a.m. EDT using the Canadarm2 on July 27. The station now had five spaceships attached: SpaceX Dragon, Northrop Grumman Cygnus, Progress 72, Soyuz MS-12 and Soyuz MS-13.

Expedition 60, Week Ending August 4, 2019

Hague opened Dragon's hatch on July 29 and he, Koch and Morgan started to unload cargo, including critical research samples, and stowing them inside the station's science freezers and incubators for analysis. Progress 72 undocked from the Pirs Docking Compartment port at 6:44 a.m. EDT. Ovchinin and Skvortsov trained on the tele-robotically operated rendezvous unit (TORU), ready for the arrival of Progress 73.

Hague and Parmitano activated the Life Sciences Glovebox on July 30 to conduct new bone research. Hague then retrieved bone cell samples to observe healing and tissue regeneration properties to promote human health on Earth and in space. Parmitano then photographed samples inside the Kubik incubator for the new Biorock space-mining study. Harnessing the power of microbes could help future astronauts extract precious minerals from the surface of the Moon and Mars. Koch and Morgan completed setting up habitats housing mice shipped aboard the SpaceX Dragon. Scientists are comparing the space rodents to a sample of mice back on Earth to understand biological changes caused by microgravity.

Progress 72 launched at 8:10 a.m. EDT on July 31 from Baikonur, carrying almost 3 tons of food, fuel and supplies. It docked at the Pirs docking compartment at 11:29 a.m. EDT.

Postscript

At the time of writing, Beyond is Luca Parmitano's current spaceflight. For his previous mission, see the Volare chapter.

A Note from the Author

In January 2019, ESA Director General Jan Wörner announced that it is his intention for all members of the ESA class of 2009 to fly to the ISS at least twice by 2024. At the time of writing (September 2019) Alexander Gerst has completed his second mission and Luca Parmitano's second mission is underway. With a Canadian mission just completed and a Japanese mission scheduled for 2020, it can be assumed that there will be an ESA long duration mission in late 2020 or early 2021. Jan Wörner has indicated that Thomas Pesquet will fly on this mission and become the third member of the 2009 class to fly twice.[1]

---

[1] At the time of writing it is looking likely that this flight will be on the first operational flight of the Boeing CST-100 Starliner. The planned crewmates are Sunita Williams and Josh Cassada.

J. O'Sullivan, *European Missions to the International Space Station*, Springer Praxis Books, https://doi.org/10.1007/978-3-030-30326-6

Bibliography

13 Dygen I Rymden, Efter 14 år på Jorden, Christer Fuglesang, Fri Tanke, 2007

An Astronaut's Guide to Life on Earth, Chris Hadfield, MacMillan, 2013

Ask an Astronaut: My Guide to Life in Space, Tim Peake, 2017

Bold They Rise, The Space Shuttle Early Years 1972–1986, David Hitt and Heather R. Smith. University of Nebraska Press, 2014

Britain and Human Space Flight, Richard Farrimond, 2013

Drum Vlucht, Het verhaal van astronaut André Kuipers, Sander Koenen. National Geographic, 2012

Elon Musk, Ashlee Vance, Virgin Books, 2015

Europe's Space Programme, to Ariane and Beyond, Brian Harvey. Springer-Praxis, 2003

Gabby: A Story of Courage and Hope, Gabrielle Giffords and Mark Kelly, Simon & Schuster, 2011

How Columbus Learnt to Fly, Glimpses of a unique space mission, Thomas Uhlig, Alexander Nitsch, Joachim Kehr. ESA, 2013

International Space Station, 1998–2011 (all stages), Owners' Workshop Manual, David Baker. Haynes, 2012

Japanes Missions to the International Space Station, Hope from the East, John O'Sullivan, 2018

Manned Spaceflight Log II-2006–2012, David J. Shayler and Micheal D. Shayler. Springer-Praxis, 2013

My Countdown, The Story Behind My Husband's Spaceflight, Lena De Winne. Apogee Prime, 2010

NASA Space Shuttle, 1981 onwards (all models), Owners' Workshop Manual, David Baker. Haynes, 2011

Praxis Manned Spaceflight Log 1961–2006, Tim Furness and David J. Shayler. Springer-Praxis, 2007

Reference Guide to the International Space Station. NASA, 2010

Ruimteschip Aarde, Ontdek Je Wereld, Met Een, Reis Door De Ruimte, Sander Koenen. Moon, 2012

Seize the Moment: The Autobiography of Britain's First Astronaut, Helen Sharman and Christopher Priest, Victor Gollancz, 1993

© Springer Nature Switzerland AG 2020

J. O'Sullivan, *European Missions to the International Space Station*,

Springer Praxis Books, https://doi.org/10.1007/978-3-030-30326-6

Soyuz, 1967 onwards (all models), Owners' Workshop Manual, David Baker. Haynes, 2014

Soyuz, A Universal Spacecraft, Rex D. Hall and David J. Shayler. Springer-Praxis, 2003

SpaceX, Making Commercial Flight a Reality, Erik Seedhouse. Springer-Praxis, 2013

The Astronaut Selection Test Book: Do You Have What it Takes for Space?, Tim Peake, 2018

The International Space Station, from Imagination to Reality, Rex Hall. British Interplanetary Society, 2002

The International Space Station, from Imagination to Reality Vol.2, Rex Hall. British Interplanetary Society, 2005

The Universe in a Mirror: The Saga of the Hubble Space Telescope and the Visionaries who Built it, Robert Zimmerman, Princeton University Press, 2010

Thomas Reiter, Leben in der Schwerelosigkeit, Hildegard Werth. Herbig, 2011

Un Passo Fuori, Umberto Guidoni. Editori Laterza, 2006

Wheels Stop, The Tragedies and Triumphs of the Space Shuttle Program 1986–2011, Rick Houston. University of Nebraska Press, 2013

Image Links

Foreword

Figure F.1:   http://www.esa.int/spaceinimages/Images/2015/02/The_European_
                Astronaut_Centre14
Figure F.2:   https://en.wikipedia.org/wiki/Walter_Peeters

Chapter 2

Figure 2.1:   https://spaceflight.nasa.gov/gallery/images/shuttle/sts-128/html/
                iss020e036847.html
Figure 2.2:   https://en.wikipedia.org/wiki/Soyuz_TMA-7#/media/File:Soyuz_
                TMA-7_spacecraft2edit1.jpg
Figure 2.3:   https://spaceflight.nasa.gov/gallery/images/station/index.html
Figure 2.4:   https://www.esa.int/spaceinimages/Images/2011/02/ATV_Johannes_
                Kepler_approaching_ISS_for_docking
Figure 2.5:   https://spaceflight.nasa.gov/gallery/images/station/index.html
Figure 2.6:   https://nasasearch.nasa.gov/search/images?affiliate=nasa&page=1
                &query=dragon
Figure 2.7:   https://www.nasa.gov/image-feature/cygnus-in-the-grips-of-the-
                canadarm2

Chapter 3

Figure 3.1:   http://www.esa.int/spaceinimages/Images/2013/04/Luca_Parmitano_
                EMU
Figure 3.2:   http://www.esa.int/spaceinimages/Images/2012/09/Soyuz_TMA-09M_
                Volare_mission_patch_2013

Figure 3.3:   http://www.esa.int/spaceinimages/Images/2013/02/Soyuz_TMA-09M_2013

Figure 3.4:   https://www.nasa.gov/mission_pages/station/expeditions/expedition36/gallery.html

Figure 3.5:   https://www.flickr.com/photos/nasa2explore/7895077550/

Figure 3.6:   http://spacefacts.de/english/flights.htm

Figure 3.7:   https://www.flickr.com/photos/nasa2explore/8144964069/

Figure 3.8:   https://www.flickr.com/photos/nasa2explore/9271250558/

Figure 3.9:   http://www.esa.int/spaceinimages/Images/2013/07/Last_smile_outside_right_before_discovering_the_leak

Figure 3.10:   https://www.airspacemag.com/daily-planet/olympic-torch-space-180949625/

Figure 3.11:   http://www.esa.int/spaceinimages/Images/2019/07/Luca_Parmitano_at_Star_City4

Chapter 4

Figure 4.1:   http://www.esa.int/spaceinimages/Images/2013/11/European_Space_Agency_astronaut_Alexander_Gerst

Figure 4.2:   http://www.esa.int/spaceinimages/Images/2013/09/Blue_Dot_mission_logo

Figure 4.3:   http://spacefacts.de/english/flights.htm

Figure 4.4:   http://www.esa.int/spaceinimages/Images/2013/09/ISS_Expedition_40_patch_2014

Figure 4.5:   http://www.esa.int/spaceinimages/Images/2013/09/ISS_Expedition_41_patch_20142

Figure 4.6:   http://www.esa.int/spaceinimages/Images/2013/11/Expedition_38_backup_crew_members

Figure 4.7:   https://www.flickr.com/photos/nasa2explore/13927928089/

Figure 4.8:   https://www.flickr.com/photos/nasa2explore/14225511769/

Figure 4.9:   https://www.nasa.gov/sites/default/files/14520703686_9cac8e27cb_k.jpg

Figure 4.10:   https://www.nasa.gov/sites/default/files/thumbnails/image/iss041e066940.jpg

Figure 4.11:   http://www.esa.int/spaceinimages/Images/2015/11/German_Chancellor_Angela_Merkel_with_ESA_s_Alexander_Gerst_and_Thomas_Reiter

Chapter 5

Figure 5.1:   http://www.esa.int/spaceinimages/Images/2014/02/Samantha_Cristoforetti

Figure 5.2:   http://www.esa.int/spaceinimages/Images/2014/01/Soyuz_TMA-15M_Futura_mission_patch_2014

Chapter 6

Chapter 7

Figure 7.10:   https://www.esa.int/spaceinimages/Images/2015/10/spacerocks_
               patch_artwork
Figure 7.11:   http://www.esa.int/spaceinimages/Images/2016/04/Road_to_
               the_stars
Figure 7.12:   https://www.nasa.gov/image-feature/the-bigelow-expandable-
               aerospace-module-beam
Figure 7.13:   http://www.esa.int/spaceinimages/Images/2018/07/ESA_
               astronaut_Tim_Peake_with_Apollo_15_astronaut_Al_Worden

Chapter 8

Figure 8.1:   http://spacefacts.de/english/flights.htm
Figure 8.2:   http://www.esa.int/spaceinimages/Images/2015/11/Proxima_
               mission_logo
Figure 8.3:   http://spacefacts.de/english/flights.htm
Figure 8.4:   https://www.flickr.com/photos/nasa2explore/25195281586/
Figure 8.5:   https://www.flickr.com/photos/nasa2explore/25118126302/
Figure 8.6:   http://spacefacts.de/english/flights.htm
Figure 8.7:   https://www.flickr.com/photos/nasa2explore/30464240742/
Figure 8.8:   https://www.flickr.com/photos/nasa2explore/33008741073/
Figure 8.9:   https://www.esa.int/spaceinimages/Images/2017/03/Shane_
               Oleg_and_Thomas
Figure 8.10:   http://www.esa.int/spaceinimages/Images/2017/06/President_
               Emmanuel_Macron_meets_Thomas_Pesquet_and_Jan_Woerner

Chapter 9

Figure 9.1:   https://spaceflight.nasa.gov/gallery/images/station/crew-26/html/
               jsc2010e045317.html
Figure 9.2:   https://www.esa.int/spaceinimages/Images/2016/11/Vita_logo
Figure 9.3:   http://spacefacts.de/english/flights.htm
Figure 9.4:   https://www.flickr.com/photos/nasa2explore/27455941791/
Figure 9.5:   https://www.flickr.com/photos/nasa2explore/36125128174/
Figure 9.6:   http://spacefacts.de/english/flights.htm
Figure 9.7:   https://www.flickr.com/photos/nasa2explore/34657061210/
Figure 9.8:   https://www.flickr.com/photos/nasa2explore/36788556772/
Figure 9.9:   http://www.esa.int/spaceinimages/Images/2017/10/Paolo_
               on_the_ISS
Figure 9.10:   https://www.cit.ie/currentnews?id=1354

Chapter 10

Figure 10.1:   https://www.esa.int/spaceinimages/Images/2017/05/Horizons_
logo
Figure 10.2:   http://spacefacts.de/english/flights.htm
Figure 10.3:   https://www.flickr.com/photos/nasa2explore/41612241712/
Figure 10.4:   https://www.flickr.com/photos/nasa2explore/44664275354/
Figure 10.5:   http://spacefacts.de/english/flights.htm
Figure 10.6:   https://www.flickr.com/photos/nasa2explore/40940977764/
Figure 10.7:   https://www.flickr.com/photos/nasa2explore/30447899707/
Figure 10.8:   https://www.nasaspaceflight.com/2018/08/soyuz-station-leak-no-
threat-repairs-continue/
Figure 10.9:   https://www.dlr.de/dlr/en/Portaldata/1/Resources/portal_
bilder/2018/2018_4/CIMON1_sn.jpg
Figure 10.10: https://www.flickr.com/photos/astro_alex/31344718597/

Chapter 11

Figure 11.1:   https://www.esa.int/spaceinimages/Images/2018/08/Soyuz_MS-
13_Beyond_mission_patch_2019
Figure 11.2:   http://spacefacts.de/english/flights.htm
Figure 11.3:   https://www.nasa.gov/image-feature/expedition-60-crew-insignia
Figure 11.4:   http://spacefacts.de/english/iss.htm
Figure 11.5:   http://spacefacts.de/english/flights.htm
Figure 11.6:   https://www.flickr.com/photos/nasa2explore/46920441404/
Figure 11.7:   http://spacefacts.de/english/iss.htm

Appendix 1

Figure A.1:   https://www.esa.int/spaceinimages/Images/2003/06/Charta_of_
the_European_Astronaut_Corps

Appendix 1

Charter of the European Astronaut Corps

© Springer Nature Switzerland AG 2020
J. O'Sullivan, *European Missions to the International Space Station*,
Springer Praxis Books, https://doi.org/10.1007/978-3-030-30326-6

Charta of the European Astronaut Corps

Our Vision

Shaping and Sharing Human Space Exploration
Through
Unity in Diversity

Our Mission

We Shape Space by bringing our European values to the preparation, support, and operation of space flights that advance peaceful human exploration.

We Share Space with the people of Europe by communicating our vision, goals, experiences, and the results of our missions.

Our Values

Sapientia: We believe that Human Space Exploration is a wise choice by and for humankind. Sapientia reflects our commitment to pursue our goals for the advancement of humanity.

Populus: We put people first, in two ways: First, the purpose of our missions is to contribute to a better future for people on Earth. Second Populus serves as a reflection of our respect for the people with whom we work: that we value their opinions, praise their work and compliment them for their support.

Audacia: We acknowledge that Spaceflight is a dangerous endeavour. While accepting the risks inherently involved in space travel we work to minimize these risks whenever we can. Audacia reminds us that the rewards will be unparalleled if we succeed.

Cultura: We continue the exploration started by our ancestors. Conscious of our history and traditions, we expand exploration into space, passing on our cultural heritage to future generations.

Exploratio: We value exploration as an opportunity to discover, to learn and, ultimately, to grow. We are convinced that humankind must embrace the challenge of peaceful human space exploration. We, the European Astronauts, are willing to take the next step.

Cologne, this fifteenth day of August twothousandone anno domini

Fig. A.1.   Charter of the European Astronaut Corps

Appendix 2

European Missions to the ISS

Mission	Astronaut	Launch Date	Launch Vehicle	Landing Date	Landing Vehicle
STS-100	Umberto Guidoni	19 April 2001	Space Shuttle Endeavour, STS-100	1 May 2001	Space Shuttle Endeavour, STS-100
Andromède	Claudie Haigneré	21 October 2001	Soyuz TM-33	31 October 2001	Soyuz TM-32
Marco Polo	Roberto Vittori	25 April 2002	Soyuz TM-34	5 May 2002	Soyuz TM-33
STS-111	Philippe Perrin	5 June 2002	Space Shuttle Endeavour, STS-111	19 June 2002	Space Shuttle Endeavour, STS-111
Odissea	Frank de Winne	30 October 2002	Soyuz TMA-1	10 November 2002	Soyuz TM-34
Cervantes	Pedro Duque	18 October 2003	Soyuz TMA-3	28 October 2003	Soyuz TMA-2
DELTA	André Kuipers	19 April 2004	Soyuz TMA-4	30 April 2004	Soyuz TMA-3
Eneide	Roberto Vittori	15 April 2005	Soyuz TMA-6	24 April 2005	Soyuz TMA-5
Astrolab	Thomas Reiter	4 July 2006	Space Shuttle Discovery, STS-121	22 December 2006	Space Shuttle Discovery, STS-116
Celsius	Christer Fuglesang	10 December 2006	Space Shuttle Discovery, STS-116	22 December 2006	Space Shuttle Discovery, STS-116
Esperia	Paolo Nespoli	23 October 2007	Space Shuttle Discovery, STS-120	7 November 2007	Space Shuttle Discovery, STS-120

© Springer Nature Switzerland AG 2020
J. O'Sullivan, *European Missions to the International Space Station*,
Springer Praxis Books, https://doi.org/10.1007/978-3-030-30326-6

Mission	Astronaut	Launch Date	Launch Vehicle	Landing Date	Landing Vehicle
Columbus I	Hans Schlegel	7 February 2008	Space Shuttle Atlantis, STS-122	20 February 2008	Space Shuttle Atlantis, STS-122
Columbus II	Léopold Eyharts	7 February 2008	Space Shuttle Atlantis, STS-122	27 March 2008	Space Shuttle Endeavour, STS-123
Oasiss	Frank de Winne	27 May 2009	Soyuz TMA-15	1 December 2009	Soyuz TMA-15
Alissé	Christer Fuglesang	29 August 2009	Space Shuttle Discovery, STS-128	12 September 2009	Space Shuttle Discovery, STS-128
Magisstra	Paolo Nespoli	15 December 2010	Soyuz TMA-20	24 May 2011	Soyuz TMA-20
DAMA	Roberto Vittori	16 May 2011	Space Shuttle Endeavour, STS-134	1 June 2011	Space Shuttle Endeavour, STS-134
Promisse	André Kuipers	21 December 2011	Soyuz TMA-3M	1 July 2012	Soyuz TMA-3M
Volare	Luca Parmitano	28 May 2013	Soyuz TMA-9M	11 November 2013	Soyuz TMA-9M
Blue Dot	Alexander Gerst	28 May 2014	Soyuz TMA-13M	10 November 2014	Soyuz TMA-13M
Futura	Samantha Cristoforetti	23 November 2014	Soyuz TMA-15M	11 June 2015	Soyuz TMA-15M
Iriss	Andreas Mogensen	2 September 2015	Soyuz TMA-18M	12 September 2015	Soyuz TMA-16M
Principia	Timothy Peake	15 December 2015	Soyuz TMA-19M	18 June 2016	Soyuz TMA-19M
Proxima	Thomas Pesquet	17 November 2016	Soyuz MS-03	02 June 2017	Soyuz MS-03
Vita	Paolo Nespoli	28 July 2017	Soyuz MS-05	14 December 2017	Soyuz MS-05
Horizons	Alexander Gerst	6 June 2018	Soyuz MS-09	20 December 2018	Soyuz MS-09
Beyond	Luca Parmitano	20 July 2019	Soyuz MS-13	February 2020	Soyuz MS-13

Appendix 3

Pre-Flight Interview with Alexander Gerst

From NASA, May 28, 2014 (Last updated 20 August 2018)

Q: Why did you want to be an astronaut?

A: I get asked a lot why I want to be an astronaut, and, you know, I don't really know when that question came up. It must have been very, very early in my life, maybe the first time I looked up in the sky and realized that out there are worlds, stars and planets, little dots in the sky that we look at that are actually real worlds, that we can fly there possibly some day and that we can find out what is out there. Is there somebody living there? What are these worlds like? If you imagine it is possible that there are more stars in the universe than there are grains of sand on Earth, that is such an amazing comparison. If you just go to a beach and pick up a handful of sand, you already have a billion grains of sand in your hand. Just imagine all the grains of sand in all the deserts and beaches of the world and that there are more stars out there in the universe than the number of these grains of sand. And we only know one of them, which is our Solar System, and we don't even know it well. So there is just so much out there that, to me, as a curious human being, the question never arose whether I would like to be part of this or not. I knew I would so early on in my life. I knew that if I ever had the chance of flying into space and to find out a little bit about our surroundings, bring a little bit of light in the darkness there, I would more than love being part of it.

Q: I want to find out about the background that led you up to that. Let me get you to start by telling me about where you are from. Tell me about your hometown and what your childhood was like.

A: I grew up in a very small town as compared to what we have here in Houston, I guess it was 12,000 inhabitants. It's a town named Künzelsau, hard to pronounce

J. O'Sullivan, *European Missions to the International Space Station*,
Springer Praxis Books, https://doi.org/10.1007/978-3-030-30326-6

in English. It is in the south of Germany and is kind of a rural town. There are forests surrounding it, and as a child if I wanted to go out playing with my friends, it was really easy as we would just go into the forest and be very close to nature. I was a boy scout and always loved being out there at night, looking up in the night sky and just being close to nature. I had an amazing family who helped my curiosity. They never tried to discourage my curiosity, even though sometimes it must have been difficult answering all those questions that this excited 4-year-old would ask about dinosaurs, storms, the sun, the stars and all that. But they somehow managed to keep up with it. They bought me books and if they didn't know the answer they just tried to find out in another way or had me read about it. I was excited to be part of this. My grandfather was an amateur radio operator. He had this little room in the basement of his house that was stuffed full with equipment. I just loved sitting there as a 5 or 6-year-old to just watch him tinker around, solder stuff, build stuff and then talk to people from different continents. Back then it was really something. It was not like we had email or could call a different continent with a telephone easily. But there, with the ham radio, you could. You just called out on the certain frequency and you got an answer from somebody from New Zealand. That was amazing to me. And actually, one day he somehow managed to adjust his antenna such that it would actually broadcast to the Moon and he gave me the microphone and let me speak a few words into the radio and the radio would actually broadcast those words as radio waves to the Moon. They would bounce back from the surface of the Moon and in about two and a half seconds later I would hear my own words in the radio knowing, as a 6-year-old, that they were just on the Moon! It was impossible for me to grasp in the beginning, to know that something of me, my voice, has just been on the surface of the Moon and now it is back on Earth. That is something that just made my curiosity even stronger, that fostered this interest I had in everything surrounding me, not just the Earth but everything that was on the surface of our Earth, underneath the surface of our Earth and out there in the big blackness above us. What is there to find? That has carried on up to the present day.

Q: It sounds like you had a lot of different interests. Your curiosity was pretty broad. Did you narrow it down to something by the time you were ready to go to the university?

A: It was always hard for me to decide what to study just because I was so interested in my surroundings. I knew it would be something related to science and engineering because I wanted to be part of exploring my surroundings and building the technology to do so, but it eventually became geophysics, just maybe from a coincidence. I was in New Zealand, backpacking after school. I was just trying to see the world, to get acquainted with different cultures and just see what is out there, to get a good base for a decision on what I would choose for my path of life. I saw the volcanoes of New Zealand that had just been active at that time and I saw the remains of eruptions and thought that was a very interesting field of science because it is new. We don't know much about volcanoes and they can even be dangerous. They affect 10% of the world's population that lives in the vicinity of volcanoes. So I thought that this is a subject where I can bring in my intuition, where I can bring in maybe new ideas and help out and then just bring some light into the darkness and that is what it turned out to be. In the end I worked on earthquakes and volcanoes, but it could have been in any other field of science that I see as relevant that I would have loved to work. I was a happy scientist by the time I got selected into the ESA [European Space Agency] astronaut corps, I was just about to organize a stay in Alaska to investigate the volcanoes there and was looking forward to that, of course. And now, to put it in the words of my colleague Samantha Cristoforetti, I swapped out the second-best job in the world for the best one and continued to work as a scientist, a little bit as an astronaut. But, of course, I have a busy day now, working with all the technology and training for flying into space and that science does not play this big a part in my life any more, for now.

Q: But it is still obviously something that is of great interest to you. Give me the high points in your professional career. What different kinds of jobs did you have and how did you ultimately end up being selected for the astronaut corps?

A: My career was basically the classical science career. I studied after school at the university in south Germany. I studied geophysics and got a diploma in Germany. I got a master's degree on a one-and-a-half-year exchange. I worked on the volcanoes in New Zealand. After that I studied in Hamburg and got my Ph.D. there. I worked on my dissertation and was working with volcanoes in Antarctica. I think that is what eventually led me towards the space program more than anything. I had been on four scientific expeditions to Antarctica. That actually is one of the places most similar to space that we can find on Earth because we work there in isolated conditions, in field groups that are consisting of two to three to five people only. We work in extreme environments, in storms, on the tips of active volcanoes at $-45°$. In a way that is quite similar to the conditions that we deal with in space. I think having worked there successfully helped the European Space Agency to make a judgment on how I would probably be able to work in space. Without knowing at that time, these expeditions to Antarctica and the education that I had acquired up to that point opened up this unique possibility that I had not even thought of before. I was dreaming of becoming an astronaut, but I knew that the chances of becoming one are so small that I didn't really plan for it. To me it was important that if I have a dream that I give it a chance one time in my life. I do not want to wake up sometime realizing that I am too old and have not tried it. So I applied to be an astronaut because I kind of told myself in the past that I would apply once in my life. I wanted to give this a solid chance, knowing, so to speak, that I would not be selected because there are too many other good people there in that program and why should I be the one? The chances are so small. So I applied in the middle of finishing my Ph.D. dissertation and actually I did not really even have time for it. I was planning for a volcanic expedition to the islands of Vanuatu, so I did not even have time to prepare for the tests or anything. I just put in my application, giving it my best effort, knowing that if I did not put in a good effort, I would be unhappy with myself afterwards. I would be in doubt. Could I have succeeded if I had put more effort in or not? I wanted to give it the best shot that I had and I was surprised when ESA actually invited me for the interviews. It continued on to a higher level in the selection campaign until we were only at a few people left and that's when, for the first time, I could not kind of keep from getting excited that I might even have a realistic chance of becoming an astronaut. In the end, when I was selected, I was surprised most of all that this one time actually did give my dream a chance to work out. So this is the lesson I learned, that whenever you have a dream, it is worth giving it a good shot, a good chance, because it might just work out.

Q: As an astronaut, when you fly in space, that part of your job carries unique risks that most of us on Earth do not ever have to confront. I am going to assume that you think those risks are worthwhile since you are doing it. What is it that we learn from flying people in space that makes it worth taking that risk?

A: When flying to space or training for a spaceflight, you are not only training for the science and the maintenance work that you do on space station. That only occupies about 10% of your skills. Most of the time you are actually training for things that could go wrong. So what are the major emergencies that we could have on the space station, like a fire or a depressurization of the atmosphere or a toxic substance that gets released into the atmosphere, things that are potentially lethal? How do we fly a spaceship to the space station? What could go wrong? What do we have to do when it goes wrong? For every single situation, what is our plan, A, B, C, D, what to do? We do that by analyzing the risk, of course. Every one of us knows that when we fly to space there is a risk that something goes wrong. We have seen this in the past with vehicles. We are sure that we have covered all the things that we could have seen going wrong, that we have improved them, but, of course, in complex systems like a spaceship or a space station you always know that there are unknown things that could go wrong, that have not been analyzed yet. So we train our skills for the best reaction to any off-nominal situations, as we call them, in space. And yet we know that the risk is there. My personal opinion about this is that if you have something that is worth a lot to you, that you think is worth doing, then it is worth taking a certain risk for that in life. To me, space exploration is definitely something that I think I can take a somewhat higher risk at for me personally because what we get out of it is so amazing. This perspective that we have, that only six people right now have up there flying in space, looking back on our Earth, being the first wave of explorers on the way out, exploring the universe to a moon, asteroids, Mars and beyond, that is for me such an important thing to do for, for humankind, for science, for international cooperation, that I think personally, for me, it is worth taking that added risk.

Q: You and your crewmates are getting set to launch to the International Space Station. Alex, please tell me about the goals of your mission and what your jobs are going to be on this flight.

A: As a crew member flying to the International Space Station you are basically doing science. This is why we have the International Space Station, to have science, to fly out to space, to explore our universe as the first step out of the atmosphere of our home planet, and to look back on to our Earth and observe it, to bring back impressions and data, scientific data, technology that we test up there and improve our life back home by flying to space. Of course, I am not only doing science on the space station. We need to keep the space station flying. It just got extended until 2024, which means from now on we have another 10 years for the space station if everything goes right. We are looking forward to keeping this treasure that we have, this international treasure on orbit where we do science, where we explore, a laboratory that has been built by more than 100,000 people from 16 nations and that is something that we value, that we have up there and that we need to keep running. So part of my job as an astronaut up there is to do maintenance, repair jobs. We might do a spacewalk to keep devices and instruments on the outside of the space station running and basically to keep a good household up there, to keep our laboratory running.

Q: Have you given any thought to what you are most looking forward to seeing when you get there for the first time?

A: Of course as an astronaut who has not flown yet, I am looking forward to being in space, just experiencing the new environment. This is something that I always was looking forward most to when I went on an expedition. Back then, when I was a geophysicist and I went to Antarctica or an exciting new place, I was always looking forward to what it feels like being there. So in space, of course, it is like how will it be to be weightless, how will it be to look out the window and see our beautiful blue planet from outside, from this unique perspective that the International Space Station gives us, that allows us to look at it, to look at the atmosphere from the outside, to see whole continents at once without any borders that we see like when we look at the at the map in school. The way we are used to seeing countries is with little lines of borders drawn around them. I am very much looking forward to seeing this from the outside without those borders, just in the way that our planet exists, without any political boundaries, just the way it looks like from space. That is something I am very much looking forward to.

Q: It has been a few years since an astronaut from Germany has been to the space station. Thomas Reiter was the last station crew member and that was back in 2006. I believe Hans Schlegel has been more recently than that. What is it like back home in Germany, the excitement about another German astronaut going and this time going for a 6-month tour?

A: Of course, Germany has a long tradition in human spaceflight. Since the early 1980s we joined the shuttle program, and early on astronauts from Germany flew to Mir and now to the International Space Station. For me it is a big honor to be able to continue this long-standing tradition and to fly to this laboratory and bring some of that excitement back with me, some of that perspective back with me, bring it home to the people in Germany. I already feel that people in Germany are excited. They traditionally like human spaceflights and they are about that. For me it is great to be part of that tradition.

Q: Your ESA colleague, Samantha Cristoforetti, is going to be arriving at the station shortly after you leave. There are going to be ESA astronauts on board for almost a full year in a row. That has got to be helping boost the excitement in Europe, too.

A: Some of the experiments that we do on the International Space Station are installed by astronauts and then mostly run by the ground so in that case it doesn't really matter which one of us installs an experiment or who runs it. But there are other experiments where we have to kind of come in with our own hands, with our intuition and, on some of those experiments, we are actually the subjects. So for these experiments like on bone loss, how muscles behave in space, how our immune system behaves in space, it is important to get long rows of test subjects so the scientists have enough statistical data to actually make conclusions and actually help life on Earth, help people who have these illnesses on Earth get treated. So in general it is very important to have this laboratory up there and have continuity in all the science experiments. It is not only for the ESA science experiments that we have on our physiology experiments side, but also for NASA it is very important to have a long row of test subjects to get a statistical base of data, and for that, of course, it is very good to have crew members fly for half a year and then not a gap but actually just like the next crew flying right after them. And for the ESA side, Samantha Cristoforetti, who will fly right after me, in fact, she will already be in Baikonur by the time that I land, she will already be preparing for her launch, and it is, of course, an important thing not only for science but also for the eyes of the public to see that we have astronauts in space continuously. We have this outpost out there. We are humans and explorers and we have our first small foot in the door to get out into space, to look at our planet from the outside, and then eventually to venture further, to the Moon, to the asteroids, to Mars. That is our first step. That is our home base that we have there, and it is important that we have this continuously manned.

Q: You mentioned a moment ago about the range of different science activities that are really going to be the focus of what you do on this mission. How do you explain to people what the potential is for what can be learned by putting people on this space station?

A: The potential for flying people in space to do experiments is huge. On our expedition alone we have about 160 experiments that we carry out in all the different fields of science that you could imagine. It is very important, for example, for experiments in the area of common illnesses that we have on Earth like osteoporosis, arterial sclerosis. If you have one of these illnesses and you are getting treatment, you are actually directly benefitting from some of the experiments that we did in space. It is a very similar condition that we have in space as astronauts where we not use our muscles and bones as to what people experience down here when they have osteoporosis where their bones get porous and get weaker. Fortunately for us we have this condition only for a while and then we can come back to Earth and recover from this condition. But it gives us the unique possibility of actually doing research on how this condition in the bones develops and how we can think of countermeasures, how we can treat it, what medication works and what does not work, because we have this accelerated situation up there on orbit. We have a lot of research in that area that helps us treat those illnesses on Earth. A scientist that I recently met actually told me that they just recently implanted a trachea in a person that they generated from stem cells that came out of space research. Some of these benefits you might not even actually hear about if you do not have one of these illnesses, but if you do then you are in a position to get much better treatment than you would have without space research. There are other areas. Technology is a big area. Several of my experiments will actually look into investigating alloys in space. It is very important, from a physics point of view, to investigate how new materials work, to investigate them in a way that you can look at them without other influences on them from the outside. In my case we actually fly an electromagnetic levitator, which is an alloy furnace that heats these alloys that we want to look at to several thousand degrees so they melt. They do not touch any vessel, any box around them, and that is the way to investigate these alloys, something that we cannot do on Earth because everything that we have eventually falls back after a few seconds to its box or something that it can be carried in. So we cannot really look at some of these physical properties on Earth, but we can in space, which allows us to calibrate computer models for future use on Earth to actually predict what are good alloys from materials that we use in future engines, for example, to use less fuel, to create less pollution or aircraft turbines of the future that might be made out of these materials. We have other furnaces up there just to look into the physical properties of how semiconductors can be made more efficient so we might have the next generation semiconductor material. Your computer in 10 years might be made out of a material that we just investigated in space. That is very exciting for me as a scientist, especially, to be part of this leading-edge technology that we investigate up there. We are in contact with the principal investigators, the lead scientists of all those top-notch experiments, and it is very exciting to actually see how they work and to work with them on an experiment. That is something that we can bring to space as humans and that is exactly where our intuition comes in. We know that in science the big discoveries were never made by looking for them. That's how science works, right? You discover things not by looking for them but you discover things on the way when you are looking for something else. That already happens on Earth and, of course, it happens in space. So these are very exciting cases where we work with the principal investigators and tell them, hey, in this experiment, something turned out a little bit different than what we expected, what should we do there; I think we should maybe repeat this investigation there or maybe change some of the initial conditions and go along a path of finding something that we did not know was there before and just find out new things. That is very exciting and I am looking forward to be part of this.

Q: I've heard the best thing to hear in the lab is not, "Oh, yes" but "That's interesting."

A: Yes, that is exactly right.

Q: On the subject of what being in that environment does to the people that are there, the station partners have decided to send a couple of crew members to space for a full year to continue that study. They are going to launch in 2015. What are your thoughts about sending a crew to space for a 12-month tour rather than six?

A: An important purpose of the International Space Station is to prepare us humans for living in space, eventually for longer periods of time and especially far away from Earth where we cannot get the help of ground, the Mission Control Center, whenever we need it, but we might be on our own out there for a long period of time. Therefore we have designed a number of experiments to help us find out more about how it is living in space autonomously for a longer period of time. One of them is basically the missions that we have that last for longer than half a year, like my mission, but last a year and maybe longer in the future. That is very valuable for us to see how the human body reacts when it is exposed to zero gravity for longer than what we have known before. Of course, we had a number of these flights that lasted for a year on space station Mir, but we need to continue this to really get a statistical database for finding out how the human body reacts out there in space. We have several experiments that I am also part of on how to autonomously work the equipment that we have on space station. Usually we have a lot of instruments on space station that we as astronauts do not really interface with because they are just not important enough for us and they can be controlled by the ground. Now if you go on the longer venture to Mars, you might be out of communication for a while, and even in the best-case communications may take 20 minutes from the word that I say until it arrives at Mission Control. Then they have to make a decision and then send it back, so it is possibly a 1 hour round trip. For these cases we have designed an experiment where I will actually operate a complex device on space station called the TOCA, Total Organic Carbon Analyzer, that is usually controlled by ground. I will work that instrument without getting specific training on it, just by using manuals or software that we have on board and by making complex decisions based on the information that we have there in real time. I will be watched by ground so they will have a look at how I do this, how I make those decisions, how we can improve that decision-making process and how it actually works to operate a complex instrument from the perspective of being alone in space with that instrument. That will give us important clues as to how we would design the next generation of spaceships that will bring us to Mars or beyond.

Q: Give me another couple of examples of experiments you are going to be working on that are about how the human body reacts to being in this environment and what you can do to try to negate the bad effects.

A: One thing we have found is that in space the immune system actually gets weaker. We don't quite know yet what that process is so we have a lot to investigate there and that is unfortunately combined with the fact that viruses and bacteria get more aggressive, become more virulent in space. That is an unfortunate combination that might get dangerous to future astronauts flying further away who might get sick. We need to investigate this and we have a whole suite of experiments that work on that topic. At the same time it allows us to actually investigate what makes bacteria or viruses dangerous in the first place and what changes the level of aggression with which they attack the immune system. That actually helps us back on Earth to create much better vaccines because we know where the target points are, where the target genes are, to attack a virus with a vaccine. We know exactly where to go because we have seen how it changed its behavior in space. So every time an organism changes its behavior, we can actually see where those changes happen and then we know this is a good target point for a vaccine. So that is how this research really helps us. For example, multi-resistant bacteria kill more people on Earth than HIV does, so it is a very relevant subject of research right now.

Q: You are also going to be participating with a lot of the countermeasures that you mentioned earlier. I have heard that some people are actually coming home fitter than when they left?

A: Yes, that's right. When we first started flying into space, space was very unhealthy for us. Just the fact that we were in weightlessness and not using our bones and muscles meant that they actually started to unbuild. They got weaker just because our body adapts so quickly to changing conditions. Our body realizes it does not need our bones and muscles so much so it starts making them smaller and weaker. Of course, that was not very fortunate when we came back from space because it could be that our muscles were too weak to carry us as we were walking. We have found out over the years how to deal with that and only recently we actually are confident enough that we can rebuild or prevent that muscle and bone loss on orbit by a number of countermeasure training systems that we have on orbit, and, in fact, some people are coming back even stronger than when they flew up there. That is a good example of how these countermeasures, like weightlifting devices and those devices specially targeting some muscle and bone areas that we have, can help people on Earth actually counteract illnesses that they have because we have initiated similar conditions to what we experience in space like bone loss, for example, with osteoporosis.

Q: You mentioned a couple of examples earlier about other kinds of science that you will be working on, where you are not the test subject, if you will, but are the lab assistant. Give me a couple more examples of what you find to be pretty interesting experiments that you are going to be a part of.

A: When people talk about the science that we do in space, they sometimes quote the number of hours that a crew spends doing science. In our case we target, for example, 35 hours for the USOS [United States Operating Segment] crew, which is the crew on the U.S.-operated side of the space station, the U.S. and ESA operate that segment together with JAXA [Japan Aerospace Exploration Agency], the Japanese space agency. We spend 35 hours a week working directly on science experiments, but actually what happens in the hours that we are not working on that is that we have maybe 10 full-time science experiments working autonomously in all the laboratories that we have. So every week we have hundreds of hours of science being performed just by experiments that we install and just leave them there. They are operated mostly by the ground, by whole teams of scientists who in control centers can remotely operate those experiments. Once in a while we may have to kind of rebuild some of those experiments, just change some parameters, restart them or, maybe something did not go as the scientists had planned. We just start something new or we take them down and preserve the data, preserve samples, bring them back down so then we invest one of these 35 hours that we spend per week on that. So really what we are up there is more like a multiplier of science. We initiate like a catalyzer of those processes and once in a while step in but most of the time science works autonomously on the space station.

Q: In some of those cases, as you said, you get to work with the scientists who dreamed up this experiment. That must be very interesting to have that kind of interactions with those people.

A: I think it is a very inspiring thing to interact with scientists who just came up with an idea that might be revolutionizing a certain field in science or technology. For example, we get to see the newest materials, some things that you might find in 10 years down the road in industrial processes, in satellites, in cars, something that really has relevant spinoffs for Earth. For example, we have a ship-tracking system installed on the space station that tries out new technologies for making sure we do not lose track of ships that travel over the ocean. So far we have a system on board that only works in the area of the coasts, but nowadays it is getting more and more important to keep track of these processes. We sometimes have piracy events on the oceans and we want to make sure we have good coverage on what is going on down on Earth, so that is a technology that we are testing on the space station without doing much as astronauts. We just install it, once in a while check if the system is working, and then if it works, we bring it back and we see that it is being used in other satellites, in commercial developments. That we have already seen this system being used is a very satisfying thing as an astronaut when you know that your work on something really has a relevant meaning down the road for other systems or other people down on Earth.

Q: From time to time station crew members have to go outside for a variety of reasons. What is the plan for spacewalks on your flight? I know things can change, but as you sit here today, what is the plan and what would you be doing during those spacewalks?

A: So far for Expedition 40 and 41 we have three spacewalks planned on the U. S. side, meaning in the U. S. EMU [extravehicular mobility unit] spacesuit, and I might be involved with one and possibly two of them. We are actually training for these so that is an exciting thing to do. One is reconfiguring some of the outside parts of the space station. We have the cooling system that is in a state that needs to be changed. We had a couple of spacewalks in the past that needed to be done because we had a few things break on the cooling system. We had the pump module that we had changed out so now we are training on reconfiguring the cooling system in its nominal way. Then we have things that just need regular maintenance outside of the space station like lights to shine on the outside of the space station for robotics operations, to make our robotic arm visible to the operator. Then we have a cable that we install on the mobile base of the U. S. robotic arm on the space station so it can move more agilely out on the outside of the space station. So it is more like the day-to-day tasks to keep the station running that you have to go outside once in a while, if we cannot do it robotically, to just keep this unique laboratory flying for the next 10 years.

Q: You look like you are a man who would be looking forward to getting the chance to go outside.

A: Oh, definitely, yes. I think going outside the door, as we say, doing a spacewalk, is probably one of the most unique things we can do as an astronaut because then you are even closer to space than you are inside the laboratory. You are in this unique, amazing spacesuit that fits like a glove and is pressurized at the same time that it protects you. Even though you do hard work, like you are in there for 7 hours and you have to work against the internal pressure of the suit that always wants to return into a neutral position, so every time you move an arm or you squeeze your glove with your hand, it feels like squeezing a tennis ball, it is a lot of hard work that you do. It is kind of like running a marathon and at the end of the day, after the training on Earth and I'm sure up there after a real spacewalk, you are actually quite fatigued. You know what you did for the last 7 hours, but, despite that, it is, I think, a unique experience that every astronaut really looks forward to.

Q: You get to go out and be your own independent spacecraft.

A: Well, that's right, the spacesuit that we use is indeed a little spaceship. It has its own life support system, a communication system, everything that we need out there to keep us healthy for 6, 7, 8 hours. It is an amazing little spaceship that I really enjoy working in.

Q: Another thing that will happen over the course of your time is that you will have visits from a small fleet of cargo ships bringing new supplies to the station. Tell me about the different vehicles that you expect you might get to see.

A: There is a small fleet of vehicles supplying the spaceship with all it needs like fuel, food, clothes, new equipment, new experiments. Some of them bring those experiments back, like the data that we get from the science that not only electronically is sent back but actual samples or specimens that we get in our laboratories will be brought back by those vehicles. There are different kinds, like the European ATV, the Automated Transfer Vehicle, which is one of the biggest cargo vehicles that automatically docks on the space station. It is a very amazing vehicle because it can do all that docking maneuver by itself, while we as astronauts just check it out. We sit in front of the camera and the computer and closely monitor what the vehicle does. We have the capability of sending it back away, but so far that has never been necessary. It was always precise to the fraction of an inch while it was docking on the space station and that will actually happen during my time up on space station if I'm lucky. I am looking forward to opening the hatch to a few fresh supplies and maybe a fresh apple, if I'm lucky, because that is, of course, a rare thing to have on space station.

Q: Apart from a European vehicle, there are others, too.

A: Yes, we have basically a small fleet of vehicles that come up there. We have the SpaceX Dragon capsule that brings us fresh supplies and actually can take down some items. Right now, apart from the Soyuz vehicle that we actually launch and land in, SpaceX is the only one that has the capability of bringing items down to Earth, so that is a very important vehicle for us because, for many scientific experiments, it is important to actually bring back specimens that we acquired on orbit. Another example that we have for a vehicle that supplies the station is Orbital [Science Corporation's] Cygnus. Then we have the Progress, the unmanned Russian supply vehicle that also flies to the space station, so we have a small fleet of vehicles that supply us. We also have the Japanese HTV [H-II Transfer Vehicle] transfer vehicle that launches from Japan right to the space station. For us these vehicles are very important. Without them we could not sustain the space station because in the Russian Soyuz, that we launch with, we only have a very small capability of launching payload and cargo with us, so we really depend on these additional vehicles.

Q: The Automated Transfer Vehicle that is due to arrive during your time on board is the last of the ATVs that is scheduled to fly. How has ESA benefited over time from the experience of building and flying these vehicles?

A: The transfer vehicle that ESA built was its means of participation in the International Space Station along with the Columbus laboratory. So it was an integral part of the cooperation between ESA and NASA to have this vehicle flying, to have it supplying the space station with a lot of items that NASA also required. For ESA it was a very useful experiment on how to build that vehicle, to test all the technologies that we needed for that vehicle to dock automatically at the space station and now the ATV-5, the vehicle that I will hopefully see docking to the space station, will be the last one of this series of five ATV vehicles. Of course, in a way it is sad to see that this is already the last one but, on the other hand, the technologies that we use for ATV are actually being used for future vehicles such as the service module of the MPCV [Multi-Purpose Crew Vehicle] Orion module that NASA builds will actually be an ESA development that is based on the ATV technology. So this is how we progress into the next generation of space vehicles. I am excited to see how that is developing.

Q: That is one good example of applying ATV technology to future missions. All of the things that you are doing, all the missions on board the space station, are designed, as you said, to help us, to prepare us, to teach us what we need to know for future missions. If you had to sum it up, tell me what it is that we are learning from these missions to the International Space Station that is critical to prepare us for the future missions of exploration.

A: I think despite all the science that we do on orbit, which is already very important for life on Earth and which really provides research possibilities that we don't have on Earth as it covers some gaps in science that we cannot close on Earth, despite all these benefits that we have from science, I think the more important reason why we fly to space and why we have the International Space Station is that humans are a species of curious explorers. Our history goes back hundreds of thousands of years, and we always were explorers. We always looked over the horizon.

What might be out there? How can we go there? How can we live there? What is the next step out there? And now compared to this several hundred-thousand-year long history of exploration, we have had the possibility to fly into space for the last 50 years. So now we are at the very start of an amazing adventure, a whole new environment, the biggest of all, arguably so, has just opened up wide in front of us, and we are taking the first steps out there. This is the most important reason why we fly into space. I think that we get the unique opportunity of taking a step out there but then not only towards the Moon, Mars, the asteroids, the universe, but actually turning back to take a look at our home planet and see how it looks from the outside. There are only six people, right now, who are outside of our home planet and can get this perspective, that can look back and bring home, soak up and bring home this perspective that they can gain from out there. From our Earth as a blue little planet, as a blue dot somewhere in space, as previous astronauts described it, with a fragile atmosphere that is very thin if you look at it from the outside — it is something that we don't realize growing up on Earth — that is easily destroyed by wars, by pollution, by everything that we do here on Earth realizing that we are actually destroying our planet. So what we can gain from human spaceflight is the perspective of how it is looking back at our planet, how fragile it is, that our planet is actually nothing but a little spaceship that flies through the universe, surrounding the Sun once a year with us as a crew. And that is exactly the point. Every one of us down here on Earth has the choice of merely being a passenger in this spaceship or being part of the crew, and I think that is an important perspective to bring back from space.

Appendix 4

Alexander Gerst Blog on Science

From ESA, November 6, 7, and 8, 2014

The Blue Dot mission is nearing its end, I leave my home of the last 5 months on Sunday night. I am touched by all who joined me on my adventure and it was a pleasure to share my experiences and photos from up here. But photography is just a hobby; the real reason the six of us are up here in space is to conduct science that cannot be done anywhere else in the world.

Although I studied and worked as a geophysicist, since I arrived on the International Space Station I have worked on experiments in the field of biology, metallurgy, physics, bacteriology, dermatology, human physiology and many more ologies! Luckily though I don't have to understand all the theory behind the experiments we perform in the enormous weightless laboratory that I have had the privilege of calling home.

In the past years, after the space station was completed, crews have constantly managed to increase the number of utilization hours, meaning the number of hours spent per week on science. Even though I hope it won't last long, our Expedition 40 crew was lucky enough to set a record for most astronaut crew time spent on research in 1 week: 82 hours on the U. S. segment alone – with a lot of help from the various control centers and the teams of scientists in the User Operation Centers. If you think this doesn't sound like a lot consider that these hours were spent totally on science and exclude eating, exercise, cleaning, operating the station and everything else you normally do in a day (granted our daily commute is brief).

In addition to the experiments that we are involved with actively, a large number of experiments run silently in the background, sending data to scientists on Earth after they have been set up. We only need to check on these experiments once in a while, and sometimes do troubleshooting.

Why do we go to all this trouble to send people to live in one of the most inhospitable places to be as a human being? Because scientists need to simplify things to understand them better.

© Springer Nature Switzerland AG 2020
J. O'Sullivan, *European Missions to the International Space Station*,
Springer Praxis Books, https://doi.org/10.1007/978-3-030-30326-6

As a geophysicist I have studied lava flow on volcanoes, but you can never be sure why lava flows as it does. Is it because of temperature, volcano shape, the ground or the type of lava itself to name just a few variables? As each volcano is different, for some research, just looking at lava in its natural conditions leaves too many variables uncontrolled so no conclusions can be made. This is why some colleagues on Earth are creating their own lava in a laboratory, to do an experiment under controlled conditions to get to the bottom of the open questions.

The science we do up here is no different, except here we can exclude the one factor that cannot be removed on Earth: gravity. Getting rid of gravity allows scientists to observe underlying forces, find constants for computer models or understand complex interactions with one less complication to worry about in their data sets. Over the course of the next 2 days I will present my favorite experiments that I have worked on, installed, participated in, monitored or simply had the pleasure of sharing orbits of Earth with.

I was very happy to turn on the Electromagnetic Levitator on Wednesday. This machine is a furnace that melts and solidifies metals with a twist. Using magnets it can suspend a metal sample in a vacuum, heat it to 2100 °C and cool it rapidly without the sample ever touching the sides of the wall. Although this sounds really hot (or cool) the machine is mainly a box that accepts cartridges to process. The whole process is automatic after we put the cartridges in, leaving us to work on other things.

I have a special connection to this experiment, as it was built in Germany and I spent many hours assembling it, whereby we overcame several hurdles. Most critical of all was a bolt that was supposed to lock and protect the hardware during its launch on the European Ariane rocket aboard ESA's ATV-5. To install the experiment, I needed to remove the bolt on orbit – but it was stuck. And even worse, it was stuck in a very inaccessible place, blocking assembly of the experiment. At first it looked like the experiment might fail, but the team of dedicated engineers on the ground and myself up here used a creative solution to fix the problem. In the end, I sawed off the stuck bolt with a hacksaw blade and shaving cream to stop metal shavings from floating into the delicate optics of the machine. It was an operation that has never been done on orbit this way, but we succeeded. This is a fine example for what human spaceflight can achieve. Had this experiment been launched on a satellite it would have failed. Needless to say, I was thrilled by our success.

Right now, this fantastic machine is doing its first test-runs. Samantha Cristoforetti will take over after I leave and continue feeding it with samples.

The Electromagnetic Levitator is a continuation of many experiments into metals in space. You might not have realized but we are well into a new space metal-age. Your smartphone probably has metal alloys that developed from research on the space shuttle in the 1980s. A few years down the road, some of the metal alloys we process in the Electromagnetic Levitator might be saving fuel in the next generation of aircraft turbines.

A more visually dramatic experiment I worked on was NASA's BASS experiment – Burning And Suppression of Solids. Basically I burnt 100 samples in ESA's contained glovebox and extinguished them. This experiment helps understand how fires spread in space – this happened on the Mir space station in 1997 – but also how flames behave on Earth, leading to safer materials and quicker fire detection.

We have also been growing crystals in space for the Japanese space agency JAXA. Using a furnace in their Kibō module we have been getting Silicon and Germanium to crystalize in weightlessness. Once researchers know more about growing these crystals they plan to apply the technology to create the next generation of solar cells and computer chips.

The International Space Station is also a great platform to launch miniature satellites or install new modules to observe our planet. In October we installed the NASA satellite-replacement RapidScat on ESA's Columbus laboratory using the Canadian-built robotic arm. RapidScat replaces a stand-alone satellite called QuikScat that proved to be essential for predicting weather. When it stopped working NASA sought to replace it quickly using the space station as ready-to-use platform. The module measures the surface of our oceans, wind speed and direction and gives early warnings for hurricanes.

NASA's B-CAT experiment is laying the foundations for using nanotechnology and will most probably increase the shelf-life of food you find in supermarkets while making food cheaper to manufacture. For this experiment we took images of colloids in space and returned them to the research teams on Earth. A colloid is the name scientists give to microscopic particles that are naturally suspended in a liquid or gas. Examples of colloids can be found everywhere from a glass of milk to shampoo. On Earth gravity is always pulling at the particles in a colloid, getting in the way of scientific observations, which is why we mixed up some colloids while orbiting Earth.

In the 19th century Italian physicist Carlo Marangoni was investigating why water and other liquids stick together slightly, a phenomenon called surface tension. Surface tension is what allows some insects to walk on water, makes balls of mercury stick together and explains why cars have droplets of water sticking to them after a car wash. Mr. Marangoni wrote about a process now called Marangoni convection that we are investigating 150 years later in more detail for a Japanese experiment in space – how cool is that? The results from this experiment will provide fundamental constants that are needed for many industrial processes. We made liquid bridges to investigate how temperatures change. The bridges are 5 cm in diameter and up to a few cm long. On Earth, gravity would pull the liquid bridges apart immediately, so we can only do this research in space.

The phenomenon behind those amazing aurora pictures I hope you have seen are particles from space hitting Earth's magnetic field. I have not only been taking pictures of this striking interaction with the magnetic field. We are investigating the influence of our planet's electromagnetic forces on electronics on the space station. This research is important as our magnetic field protects us from cosmic rays that can crash computers and even aircraft. As aircraft and humans fly higher and higher (not to mention satellites) the electronics used are more susceptible to hardware failure from these cosmic rays. The auroras are just a beautiful bonus.

The last blog entry on my favorite science experiments will focus on biology and medicine. Going to space has changed my body in many ways. I receive more radiation up here, since we do not have our planet's atmosphere to stop harmful solar and cosmic particles. My spine has enlarged due to living in weightlessness and the blood in my body does not naturally fall to my feet as on Earth.

A human body is amazingly resilient; I hardly notice any of these changes, and a team of medical doctors monitors all astronauts to make sure we stay healthy. Regardless, these factors put a strain on our bodies that is similar to rapid aging. If we did not exercise, our bones and muscles would waste away. Our skin is less elastic, not to mention that our immune system works less well. Research teams have been looking at ways to prevent these changes since the early days of spaceflight, and great advancements have been made. Some of my muscles have surely shrunk since I arrived up here, while others have even grown since I started working in the space station. But then I do work out 2.5 hours each day up here – doctor's orders.

Much of the research in medicine and biology is looking at helping astronauts preparing for even longer missions further away from our planet, but the results are directly applicable to people on Earth. Having six people living in close quarters for 6 months in a controlled but extreme environment is an opportunity for many researchers to study changes to our bodies.

We collect samples of bacteria on and in our bodies, we record our immune system state through regular blood, urine and saliva samples, play with different diets to see the effects, have our brains scanned and do ultrasound on our hearts. All this research is helping medical researchers understand in more detail how our bodies cope with living in stressful conditions and how we age.

This type of research on the space station has already developed new vaccines against salmonella, helped asthma patients monitor their affliction, improved guidelines to combat osteoporosis and arteriosclerosis and hinted at the mechanisms we could use to keep human cells living healthily forever.

One experiment in particular could have far-reaching consequences. We are investigating an approved drug used against diabetes that also appears to stop cancers from developing. Using the drug on yeast cells up here will show how it works against cancer and allow researchers to adapt it for use on Earth. We do this research in space because weightlessness appears to affect drug delivery in yeast cells on a molecular level in a way that resembles how a human body absorbs medicine. The weightless yeast is a stand-in for human testing.

Reid and I have been looking closely at our skin for ESA's Skin-B experiment. We collected data with a number of machines to check the structure, oxygenation, hydration and elasticity so that scientists on Earth can develop models of how skin ages. This could be done on Earth, but it would take decades to monitor people throughout their lives, while our skin provides valuable information more quickly. Luckily, most of the effects are reversible after we come back on Earth, which is yet another opportunity to study how aging works.

Throughout my mission we have been growing plants in space to prepare for even longer missions, where supplies from Earth are not possible. Fresh food is one of the things astronauts miss most while in space – there were times up here when I would have traded a kingdom for a bowl of salad. So you can imagine just how much I look forward to eating fresh fruit and vegetables soon!

The research done to understand how plants grow and improve their yield is of course great for our planet, which is supporting more and more people every year. I have sent many pictures of agriculture in water-scarce areas, and with the current climate change this problem will get more and more severe for certain areas on our planet. Our research might eventually help to grow plants more water-efficiently, meaning to produce more food with less water.

Right next to my sleep station, in the KIBŌ module, a Japanese experiment is growing protein crystals without gravity, which scientists think could lead to a whole new class of pharmaceutical drugs. Our body has over 100,000 kinds of protein, and we need to understand how they work, because they are involved in almost every function in our body – from DNA to transporting molecules. Unlike on Earth, in weightlessness we can grow perfectly formed crystals that allow researchers to pinpoint the detailed effects of chemical reactions, while knowing that gravity had nothing to do with the final shape. Once we know how and when proteins are active we could apply substances that stop a protein from function-ing – halting a disease in its tracks!

If you followed my mission you might have seen a few pictures of an orbital sunrise. We experience 16 of these every day, and each and every one of them appears to be more beautiful than the last. On the space station we live according to UTC time, but nobody really knows how our bodies adapt to living in a place with no real night and day cycle. To get behind that, I have been wearing a tem-perature sensor and recording levels of melatonin to monitor my sleepiness and alertness in an experiment called "Circadian Rhythm." This research will have direct benefits for people that work long or irregular hours – emergency doctors, ambulance drivers or night-shift workers.

A more fun experiment (for me) was driving the Meteron robot in the Netherlands using a new 'space internet.' This experiment is designed to control robots on a distant planet, but the technology behind it is perfect for remotely operating in areas that lack infrastructure, and where precision and dependability is vital such as for earthquake response or remote-controlled surgery.

Not all the experiments we do here are about improving lives. Some are actually investigating where life on Earth came from! Pay a thought to the bold organisms on ESA's EXPOSE facility that are right now being subjected to outer space. Strapped to the outside of the International Space Station these bacteria, cells and organic compounds have to endure the freezing cold of –12 °C to +40 °C, the blinding sun, cosmic radiation and a deep vacuum for 18 months. To humans, this environment is so deadly that it took Reid and me 4 hours just to dress up and get ready for our spacewalk. In contrast, these organisms are out there entirely unprotected – and yet they survive. Amazingly, previous EXPOSE experiments have shown two types of organisms that can survive in outer space: lichen and tardigrades, also called "water bears."

This research has proven that it is possible that life could have traveled to Earth from another region in the universe without the need for a spacecraft. Could comets and asteroids possibly serve as a shuttle to distribute life forms in the universe? That is one of the next big mysteries for space research to solve. Stay tuned! Two days after Reid, Max and I land on Earth, ESA's comet hunter Rosetta will investigate this question when she sends her Philae lander to explore a comet from up close for the very first time – right on the surface of comet 67P. While I am getting accustomed to gravity again at the European Astronaut Centre in Cologne, I will definitely be watching this amazing venture of Rosetta.

Appendix 5

Tim Peake's FAQ

From ESA March 27, 2016

Note: Questions 20–37 added June 6, 2016. Question 38 added October 25, 2016.

Thanks for all your comments and questions on social media, your interest is just fantastic. I've answered some of the most commonly asked questions, and listed them here. I'll try to answer more, but before posting new questions, please check through these to make sure I've not already covered that topic.

1. **What time zone do we use in space?**
   The space station runs to UTC (Universal Time Coordinated), which is basically GMT, luckily for ESA and those of you in UK. This time zone was chosen because it's 'in the middle' of all the International Space Station partners (USA, Canada, ESA, Russia and Japan) and each of the two main mission control centers, Houston and Moscow, gets to cover half a day shift.

2. **Do we all sleep at the same time, and what time do we go to bed?**
   Yes, we sleep pretty much all at the same time. We try to get about 8 hours of sleep per night, but this varies, and we can go to bed anytime from 10 p.m. to midnight. I usually wake up at 6:30 a.m., which gives me time to prepare for the day's activities. While we're sleeping, mission control centers are monitoring the ISS systems. In case of an emergency, the ISS alarm would wake the crew, and the station commander also has a radio speaker in the crew quarters.

3. **Is there gravity in space, and why do we appear 'weightless'?**
   This is a very common misconception that there is no gravity in space. Gravity is everywhere in space! It's what keeps the Moon in orbit around Earth, it keeps Earth in orbit about the Sun and holds galaxies together. So, why doesn't the space station or satellites fall to Earth, and why do we appear to be floating

in the space station? That is because of our speed. We are not floating, we are actually 'falling.' But we do not fall to Earth; because of our high speed, we fall 'around' Earth. We travel about 28,000 km/h (17,500 mph), but as we accelerate towards Earth under the pull of gravity, Earth curves away beneath us and we never get any closer. Since we have the same acceleration as the space station, we feel 'weightless.'

4. **What does it feel like to float in space?**
   Floating in space is the most incredible feeling. It's actually very liberating not to be held down by Earth's gravity. I can look at something in the ISS 'ceiling' and just pop up there, turn upside down, pick it up, do a somersault and come back down again. We can store items on the walls and overhead without worrying about them falling down (Velcro is your friend here). It takes a while to get used to weightlessness and moving efficiently around the station – but once mastered you can move around quickly with just the smallest of pushes. Imagine being able to travel down a long corridor on Earth without any effort – it would be great! We hardly use the soles of our feet at all – and because of this they become very smooth and soft – but we are constantly using the tops of our feet to grab under handrails and hold us down and so the skin becomes rough and hard on the tops of our toes.

5. **How do we wash our clothes in space?**
   We do not have a washing machine, so we wear the same clothes, including underwear, for several days before we change. It is not as bad as it sounds. We live in a temperature-controlled environment, so clothes do not get as dirty as they might on Earth. Some of the items, like socks or our exercise equipment, have anti-bacterial materials in them. We change our exercise kit every 7 days. We change underwear every few days. Then we have polo shirts, T-shirts, trousers and shorts that last weeks. I'm still wearing the same pair of trousers I've worn since day 1... hmm maybe it's time for a change!

6. **How do we cut our hair and shave in space?**
   We use a set of standard hair clippers for cutting hair. The only modification is that it is connected to a vacuum cleaner to suck up all the hair. I've been cutting my own hair just for fun and it's working out OK, but not something I will continue to do on Earth! For shaving, we use either electric razors, or regular razors with shaving cream. In using a regular razor, though, you have to wipe the shaving cream off on a piece of tissue quite often. When using the electric, we shave in front of an air filter, and this catches the stray whiskers! I have a wet shave at the weekends and use an electric during the week.

7. **What happens to waste from the Space Station?**
   Our used and dirty clothes are placed in a waste bag and put on a supply spacecraft (Progress or Cygnus) that undocks and then burns up in Earth's

atmosphere. This is also how we get rid of other rubbish, such as empty food packaging or our solid waste from the toilet, for example. We don't get rid of urine – that is recycled back into drinking water.

8. **Can I see stars and planets, and do they look different?**
   Yes, we can see stars and planets. We still see some twinkling of stars because of the vast amount of gas/dust in the universe that starlight passes through. However, our atmosphere causes most of that 'twinkling' and makes the stars appear slightly less clear than when viewed from space. That's why you find many of the world's observatories on mountains tops – it's an attempt to reduce the amount of atmosphere that starlight has to travel through (and mountains tend to have less light pollution!).

   Of course, we have no atmosphere outside our window, so it seems to me that the planets appear to be slightly brighter when viewed from up here – certainly Jupiter, Mars and Venus. What's also interesting is trying to judge distance. When Cygnus departed in February, we had the most spectacular view of it disappearing ahead of the space station and of course it got smaller as it got further away, but it still looked very sharp and clear even at a distance, because there were no atmospheric obscurants in between. This made it harder to judge how far away it was.

9. **Why are there no stars in my pics of Earth or my spacewalk?**
   The reason why you can't see stars in my daytime pics of Earth or my space-walk is that, when lit by the Sun, any foreground objects, such as Earth, the space station or my spacesuit, are many thousands of times brighter than the stars in the background. Earth is so bright that it swamps out most if not all of the stars. The stars don't show up because the camera cannot gather enough of their light in a short exposure. Our eyes are a lot more sensitive to light than the cameras. To take pictures of stars, we need longer exposures on night-time parts of our orbit.

10. **How many times will we go around Earth during our flight?**
    You can work this out for yourselves. First, you need to know how long we'll be in space. Let's say 170 days. Then you have to know that we have 16 orbits every day. So that's 170 × 16 = 2,720 orbits.

11. **How far will we travel during our time in space?**
    You know that we travel at 28,000 km/h. Then you need to work out how many hours there are in 170 days. The distance will be 170 × 24 × 28,000 = 114,240,000 km.

12. **How long does it take to get to the Space Station?**
    It takes a thrilling 8 minutes and 48 seconds to be launched into space on our Soyuz rocket. Once in space we have to do a few things. First, we have to

'normalize' our orbit because it will be elliptical after insertion, and we need to make it more circular. Then we need to raise our height up to about 400 km using something called a 'Hohmann transfer' and then we need to put the brakes on so that we match the ISS speed and height perfectly in time for docking. This all takes about 6 hours from launch.

13. **Why don't you need a heat shield when leaving Earth, but only on re-entry?**
    During launch, we gain altitude at relatively low velocity at first, but by the time we hit the high Mach numbers, we're already in space, so there's no air to cause frictional heating. During re-entry, we have no choice but to use the atmosphere to slow us down, all the way from Mach 25 (28,000 km/h), and that creates a lot of heat.

14. **Did you always want to be an astronaut?**
    As a young boy I was always fascinated by the stars and the universe. When it came to a career choice, though, I was passionate about flying and could not wait to train as a pilot. I spent 18 years flying all types of helicopters and aircraft and eventually trained as a test pilot. I followed the space program closely, and when the European Space Agency held its selection for astronauts in 2008 I was ideally placed to apply.

15. **What food do you eat in space?**
    We eat fairly normal food, like you might eat on Earth, but it is out of cans or packets. Some of it is dried food to which we add water to make it edible. Other (irradiated) food comes in pouches which we place in our electrical food heater to warm up. It's not too bad, but we try not to eat too much salt in space, as it can exacerbate the loss of bone density and so the food can sometimes be a bit bland. The portions are also quite small so you have to be careful not to lose too much weight… a great excuse for eating dessert every night! My favorite foods are the breakfast menus (scrambled eggs, baked beans and sausages!). We also get a very small supply of fresh fruit every so often on the supply spacecraft.

16. **How can you eat food in weightlessness? Doesn't it float back up?**
    You can swallow and digest food just fine in space. Digestion relies on different muscles in your body that squeeze behind the food you eat, pushing it through you until it's safely in your stomach. This is called 'peristalsis.' Once the food is in your stomach, you have various valves that keep food in place. However, after a meal you can definitely feel that your food sits more 'lightly' in your stomach and I learned the hard way that you shouldn't run on the treadmill for at least an hour or two after eating!

17. Is it true that you lose your appetite in space?

Not really, but the sense of smell is incredibly important when it comes to both generating a desire to eat and tasting food. In weightlessness, food aromas don't hang around in the same way they do on Earth. Salt and pepper (which are suspended in liquid to stop the particles floating away), and other condiments are used to enhance the flavor. But our ability to taste food is affected in another way. Because fluids inside us shift towards the center of our bodies and, in particular, our heads, we get tend to feel like we have stuffy noses, making it more difficult to smell food.

18. What would happen if someone got sick or injured in space?

All astronauts are trained to a very high level in first aid. In addition, there are always at least two Crew Medical Officers (CMOs) on board that can deal with basic surgical procedures, such as filling teeth or suturing, for example. Both Tim Kopra and I are trained CMOs. We also have a medicine cabinet, which is like a small pharmacy, containing everything from analgesic painkillers and antihistamines to sleep aids, all the way up to antibiotics and local anaesthetics. We also have an Automated External Defibrillator (AED) on board for resuscitation, and we have procedures that include using a spacesuit as a personal pressure chamber if we needed to treat a crew member suffering from 'the bends' following a spacewalk. If we developed a serious illness, like appendicitis, for example, the situation would be assessed by our doctors on the ground whether it would be better to stay on board and use medication, or to return to Earth (appendix removal is not standard procedure for astronauts prior to flight).

19. What would happen if there was a fire on the station?

We've already had a couple of emergency fire warnings since I have been on board, but thankfully both turned out to be false alarms. However, we treat every situation as a real emergency, of course. Fires can vary between open fires (visible flame), smoke, smell of burning, or just a smoke detector alarm, but no other indications. We have procedures that deal with each case depending on the severity of the situation. In the most serious cases, we would don breathing apparatus and fight the fire using either carbon dioxide, water mist or foam fire extinguishers. We would also try and locate the power source and remove electrical power (electricity is most likely to be the cause of a fire on board). The smoke detectors trigger an automatic response from the ISS to shut down all ventilation systems, so as not to feed oxygen to the fire and to reduce the spread of smoke throughout the station. We have special detectors that tell us if the air is contaminated with carbon monoxide and other harmful gases – these are used to know when it is safe to remove our breathing masks. We also have special filters and equipment on board to clean and scrub the atmosphere to return the ISS back to full health. Only as a last resort would we evacuate the ISS in our Soyuz spacecraft. Interestingly, the Soyuz does not have any fire extinguishers. The way to fight a fire in the Soyuz is to close your helmet and depressurize the whole spacecraft… no oxygen = no fire!

20. **With the space station moving so fast, do you ever feel motion sickness?**

We do move quickly across Earth's surface (ten times the speed of a bullet… that's pretty quick!), but this visual effect does not cause any motion sickness. Of course, it takes a short while to adjust to microgravity, and during the first couple of days in space you might feel some 'space sickness.' This is different from motion sickness, though. We normally associate motion sickness with long periods of feeling unwell, and it can be quite debilitating. Space sickness on the other hand can come and go quite quickly, even catching you by surprise, and it's possible to function normally in between short periods of feeling unwell. Once our body learns to 'accept' this conflict between vestibular and visual information, we can do things in space that would make you feel very sick on Earth. For example, I have tried spinning in space for several minutes at a rapid speed while moving my head up and down with my eyes open. This kind of stimulation on Earth would be very provocative – but in space it's no big deal. I can sometimes make myself feel dizzy, but there is no associated nausea. During my spacewalk there was one point when I was returning with the failed 'SSU' that Tim Kopra and I had replaced. I had to descend along a spur that connects the main truss to the airlock. Half way along this spur I looked down at Australia beneath me and felt a sudden feeling of vertigo. It made me smile because I had been 'out the door' for well over an hour by that stage and it caught me by surprise. NASA astronaut Chris Cassidy had told me that, if that ever happens, wiggle your toes and it will make you relax your grip… it worked!

21. **In scuba diving, there is a syndrome known as 'fear of surfacing,' where divers don't want to come up. Did you ever feel like this on your spacewalk?**

I've also enjoyed some spectacular dives where the temptation to go deeper and stay longer is strong. As for coming in at the end though – I think the circumstances of our spacewalk made it very clear it was time to come in. After all, you can clear your mask easily under water – it's a bit harder to clear a helmet filling with water in space! Also, our spacewalks are choreographed to the minute, and we have a huge support team in Mission Control looking after us and checking our every move, so there's always someone ready to remind us of when it's time to come back in, no matter how tempting it may be to stay outside.

22. **Can you see aircraft or ships from space?**

It's not easy to see small objects with the naked eye – often you can pick out a ship's wake or aircraft contrails and follow it back – but with zoom lenses, we can see large ships.

23. **In the photos of the aurora, is this how they appear to the naked eye or are the colors more visible because of the camera exposure?**

    For the aurora photos, they are pretty close in terms of color and intensity – but if anything, we see a more spectacular image with the naked eye. The camera doesn't do justice to the eerie way the aurora snakes, ripples and changes intensity – although some of the time-lapse sequences have captured this.

24. **You've trained for most situations on the ISS, but what thing surprised you most when you first got into space?**

    The thing that most surprised me was how black space appears during the day. You know that stars are out there, but because your eyes adjust for brighter objects, space looks so incredibly dark. During my spacewalk, it felt almost intimidating being on the farthest edge of the space station and having nothing but the vast blackness of space over my right shoulder. That was a good incentive to not let go!

25. **What feature or place on Earth's surface do you look forward most to seeing on each pass?**

    Well, on viewing Earth's surface, it's all truly amazing. Often I go to the window expecting to see a certain mountain range, city or other landmark, but I'll come away with photographs of something completely different. Earth has so many secrets and the longer you spend in space the more time you have to find and appreciate them. Even every sunrise and sunset is unique and special in its own way.

26. **Do you dream differently in space, or dream of anything in particular?**

    Not particularly. I think I sleep more lightly in space and don't dream that often – when I do, I have been on Earth.

27. **What's your favorite part of your day in space?**

    I enjoy wrapping up at the end of the day and having some time to take photographs, look out the window and call friends and family. The working day is fun, but we're always trying to keep to a tight timeline, so it's nice to have a couple of free hours in the evening.

28. **Do you have any personal reading material, and what would be your choice of book to read in space?**

    It's here, I have an original edition of Yuri Gagarin's autobiography *Road to the Stars*, this is Helen Sharman's personal copy, signed by her crew and Gagarin himself. I can't think of a better choice, and it's an honor to borrow this book and read it up here.

29. How do you tweet and Facebook from space?
Both great questions. I'm tweeting and posting from a regular laptop inside my crew quarters. Our signal is relayed via satellite to a desktop PC at Mission Control in Houston that mirrors what we do on our screens up here (for security purposes, there's no direct Internet connection). It's much slower than your average wifi (think dial up speed!) and only available at certain times during the day, depending on satellite coverage. However, the fact that we get Internet at all in space is quite remarkable, and it's OK for basic access to social media and for reading news websites – but it doesn't provide the bandwidth for video streaming. So yes – I do see your comments, and although I'm not always able to reply to all, I do appreciate them very much!

30. What camera do I use?
All my Earth pics have been with a Nikon D4 and one of these lenses.

31. What kind of watch do I use?
All ESA astronauts are issued with the Omega Speedmaster X33 Skywalker, more info in this ESA article.

32. Can we see other planets from space?
Yes, we can see other planets from space, and I have been able to photograph Venus rising over Earth and also Jupiter, Mars and Saturn. Most of our windows look down on Earth – so although we see the planets rising and setting it's much harder to see them when they're above us. Here's Venus rising just before the Sun (https://flic.kr/p/GXKsxZ).

33. What's it like to sleep in space, and where do I sleep?
I sleep in my crew quarter, which is a bit smaller than a public telephone box, but big enough for everything I need. We try to get about 8 hours of sleep per night, but this varies, and we can go to bed anytime from 10 p.m. to midnight (we're on GMT). I strap my sleeping bag loosely to the wall and then zip myself into it and let myself float. Our sleeping bags are quite close fitting, which is good because you don't want to move around inside them too much. I find it easy to sleep in space, but it's probably not as good quality sleep as I get on Earth. It's sometimes hard to get your arms into a comfortable position – I normally just fold them across my chest. I am actually looking forward to sleeping in a proper bed again and having the feeling of gravity pull me down into a comfy mattress!

34. What's my favorite space food?
I love making myself a peanut butter and jam sandwich in the afternoon. We don't have proper bread so I use a tortilla wrap instead, but it tastes pretty good! My favorite foods are the breakfast menus (scrambled eggs, baked

beans and sausages!). Also, occasionally we have 'Maple Muffin Pancackes,' which taste great for breakfast – especially with a bit of extra honey on them! And, of course, I have my favorite 'space dinners' salmon dish from the kid's competition and designed by Heston Blumenthal.

35. When do you come home? What happens after you've landed?

We undock and return home June 18. Undocking will be 05:46 GMT, and we'll land around 09:15 GMT in Kazakhstan.

36. How fast is your re-entry, and what does it feel like?

Well, I've only done simulations so I can't tell you from experience (yet!). After undocking, the Soyuz has to move away from the station at a careful 0.1 m/s. Then there's a 15 seconds separation burn of our engine when the Soyuz is about 20 m from the station. When we're about 12 km from the station, our main engine fires for 4 minutes 38 seconds to slow us down. This reduces our orbital speed of 28,800 km/h and sets us on course to enter Earth's atmosphere. (After this deorbit burn we're coming back to Earth whether we like it or not!)

Once we've completed this burn, we no longer need the Service Module and Orbital Module, so pyrotechnic bolts fire and they separate and burn up on re-entry. Our Descent Module will be tumbling slowly as it enters Earth's atmosphere at about 120 km altitude. This is where we get the greatest heating effect, and our speed is dramatically reduced.

Aerodynamics start to take effect, and the capsule orients itself to enter the atmosphere, heatshield first. We'll be pushed back into our seats with a force of 4 to 5 g for several minutes while the heat shield protects us from temperatures of up to 1600 °C. This means we feel four to five times our own body weight, and during this time the plasma around our spacecraft will prevent us from communications with mission control and the ground search and rescue team.

Next we will feel the jolts of our drogue and main parachutes deploying. These slow us down further, the drogue from over 800 km/h to 250 km/h, and the main to 25 km/h. This is a very dynamic event – the capsule will be rotating quite violently beneath the drogue chute as it slows us down rapidly. Once the main parachute opens, we'll have about 15 minutes until landing. During this time we'll establish communications with the search and rescue team and prepare the spacecraft for impact.

At the last moment before touchdown, 'soft-landing thrusters' fire to limit the landing speed to around 5 km/h. We sit in custom-fitted seats that absorb some of the shock of impact – but something that all astronauts agree on is there is no such thing as a 'soft landing' in a Soyuz. We're advised to tighten straps as much as possible, push your neck firmly back into your seat, brace arms and knees together tightly and make sure your tongue is not between your teeth! Finally comes the hatch opening and the first fresh scents of Earth since leaving the planet in December.

37. What happens after you've landed?

After leaving the Soyuz, we'll fly by helicopter to Kostenay, from where Yuri will continue to Star City near Moscow, and Tim and I board a NASA aircraft to Bodo, Norway. As the aircraft refuels, I'll say goodbye to Tim and transfer to another plane to Cologne in Germany. I'll arrive at the European Astronaut Centre (EAC) in the early hours the next day. This is the home base of all ESA astronauts.

Here, ESA's medical team will supervise my rehabilitation after spending over 6 months living in weightlessness. In addition to rehab, I'll have many medical checks related to the human physiological science studies that I'm participating in and debriefs on various aspects of my mission.

The first couple of days on Earth will probably be quite tough as my bones and muscles adjust to coping with gravity once again. In addition, my balance will be a bit messed up since my brain has sort of switched off the signals coming from my vestibular system, as these are not required in space and only contribute to space sickness. However, on Earth that shifting fluid in our inner ear is vital for helping the brain to detect the gravity vector and to maintain our balance accordingly.

Also, I'll take some salt tablets and try and load up on fluids during the final few days in space, since my body spent the first month or so getting rid of the excess body fluid that is not required in space, but I'll need that back again on Earth. Lack of body fluid, low blood pressure and a weak cardiovascular system can contribute to 'orthostatic intolerance' (aka fainting) or light-headedness when standing up too quickly.

38. Who has been your hero or inspired you in your career?

For me, one of the most inspiring figures has been NASA astronaut Bruce McCandless, the first astronaut ever to do an untethered spacewalk. Just because what he did was really ground-breaking: a very high-risk activity and the level of isolation he must have felt hundreds of meters from the space shuttle. Just him and a jetpack, no cables, no tethers. It takes a lot of guts to do that. Plus he was selected back in the 1960s as one of the original Apollo astronauts, but he had to wait 18 years to make his first flight. Eventually, he logged over 312 hours in space, including 4 hours of EVA jetpack flight time. Just shows if you stay focused on your dreams what you can achieve.

Appendix 6

Alexander Gerst FAQ

From ESA, November 20, 2018

1. **How long does it take to get to the International Space Station?**
   Traveling from the launch site in Baikonur, Kazakhstan to the International Space Station can take anywhere from 6 to 48 hours depending on launch procedures and the station's position in orbit.

   ESA astronaut Alexander Gerst launched into space alongside NASA astronaut Serena Auñón-Chancellor and Roscosmos commander Sergei Prokopyev on board the Soyuz MS-09 spacecraft at 11:12 GMT (13:12 CEST) on Wednesday June 6, 2018. After 2 days and 34 orbits of Earth, the trio arrived safely at the International Space Station as planned on June 8 at 13:01 GMT (15:01 CEST).

2. **How high and at what speed does the International Space Station travel?**
   The International Space Station circles Earth once every 90 minutes. It travels at a speed of around 28,800 km/h and an altitude of 400 km above Earth. Use this map to see where the space station is, along with its exact speed and altitude.

3. **Can I see the space station from Earth and how?**
   Yes! In fact, many of you already have. As the third brightest object in the sky, the International Space Station looks a lot like a plane – only traveling much higher and much, much faster.

   NASA site Spot the Station provides important information on sighting opportunities in your local time zone, including: the maximum time period during which the station will be visible, its height from the horizon in the night sky, the location in the sky where it will be visible first and the point at which it will disappear.

   You can also use an app like ISS Spotter or check out this tracker on the ESA website.

© Springer Nature Switzerland AG 2020   312
J. O'Sullivan, *European Missions to the International Space Station*,
Springer Praxis Books, https://doi.org/10.1007/978-3-030-30326-6

4. **Is there gravity in space?**

Short answer: yes. Gravity, or rather microgravity, is everywhere in space. It is what keeps the Moon in orbit around Earth and Earth in orbit around the Sun. So why doesn't the space station fall back to Earth, and why do astronauts appear to be floating in space? That's all to do with speed.

In truth, astronauts are not floating, they're 'falling.' But because of the high speed of the space station's orbit, they fall 'around' Earth – matching the way Earth's surface curves. The Moon stays in orbit around Earth for this same reason. Since astronauts have the same acceleration as the space station, they appear and feel weightless.

5. **What does it feel like to float in space?**

In his first blog post from the Horizons mission Alexander said: "When the hatch opened to the station, it felt like coming home to me. I immediately got along as well with floating as a fish does with water. Interestingly, my feet and hands have even unconsciously remembered where handrails and foot loops are, and usually, when I flew around a corner, they grasped the correct handrail on their own. But there was one time it didn't work out so well. My body thought it knew exactly where a handle had to be and reached out to grab it – only to find someone had moved it since my last mission…"

In this earlier interview about his Blue Dot mission, Alexander said that in the beginning life on board the space station was a lot like jumping from a 10-meter-high diving board. While he didn't get sick, he did experience a tingling in his stomach for the first 2 hours, and it felt like he was hanging from the ceiling like a bat.

After 2 days, his body adapted to the point where it all just seemed normal.

6. **What is the temperature on the space station?**

The temperature inside the International Space Station is kept constant at around 22 °C.

7. **What time is it on the space station?**

The International Space Station uses UTC, or Universal Coordinated Time. This is also referred to as GMT or Greenwich Mean Time and is 1 hour behind CET (Central European Time).

UTC was selected as the time zone for the station as it is around the midpoint for all International Space Station partners and allows the two main mission control centers (Moscow and Houston) to cover half a day shift each.

8. **What kind of watches do astronauts wear?**

ESA astronauts wear an Omega Speedmaster X33 Skywalker – a new version of the historic space watch, tested and qualified with ESA's help.

The design of this watch draws on the inventions of ESA astronaut Jean-François Clervoy, who flew to space three times in the 1990s. You can read more about this watch's development here.

Note: ESA is an intergovernmental organization and is not involved in the manufacturing or commercialization of the Omega Skywalker X-33.

9. **What food do you eat in space?**

According to Alexander, food on board the International Space Station "tastes better than it looks." That's just as well as it needs to sustain astronauts through a busy schedule of science and operations.

Most food comes in cans or packets. Some is dried, and astronauts add water to this to make it edible. Other (irradiated) food comes in pouches that astronauts place in an electrical food heater to warm up.

This post on the tastes of space explains more about the crumb-free, lightweight and preservative-free criteria space food must adhere to. You may have also seen Alexander's posts about growing salad in space. This kind of activity will be essential for longer term deep space missions.

10. **How do astronauts wash their clothes in space?**

There is no washing machine on the International Space Station, so astronauts must wear their clothes for several days before changing them. Luckily, in the space station environment these don't get as dirty as they might on Earth, and some items such as socks and exercise equipment contain anti-bacterial materials to help prolong their wear.

Exercise equipment is changed weekly, underwear is changed every few days, while shirts and pants can be worn for weeks.

11. **What is an astronaut's daily routine?**

An average day in the life of an astronaut on the International Space Station looks a little bit like the below:

6:30: Wake up and start the day with the space version of a shower using wet cloths and dry shampoo. Brush teeth with edible toothpaste.

7:00: Eat breakfast. Alone or with other crew mates. There is coffee and tea, but instead of bread or bread rolls (that might crumble) astronauts instead choose tortillas or other space-suitable breakfast options.

7:30: Daily conference with ground control to review the day's tasks and answer any questions. One after the other the astronauts are connected with relevant ground control stations in the United States, Russia, Europe and Japan. This takes 15–30 minutes.

7:45: Start of actual working day. Astronauts spend the majority of their time working on scientific experiments according to that day's schedule.

Between 12:00 and 14:00: Lunch break. Alone or with other crew members.

After lunch: More scientific experiments, exercise and other scheduled tasks.

Between 19:00 and 19:30: Evening conference. Every ground station takes part in this. The astronauts report on the experiments they conducted and any issues they may have had. This evening catch-up is usually very quick.

19:30: End of work day. Now astronauts can write emails, read, listen to music, watch movies or take photos of Earth from the Cupola window.

20:00: Dinner with International Space Station crew

22:15: Bed time. Every astronaut decides when to go to bed. While sleeping, astronauts wear earplugs because it is relatively loud on station.

12. **What camera equipment does Alexander use?**
   Alexander takes some fantastic photos both inside the International Space Station and of Earth below. In fact, imagery captured on board the station has progressed from 35-mm film and SD 4 × 3 video to 20.8 MP digital stills and 8K video in the 20 years of space station operation.

   Most of Alexander's current shots are taken with a Nikon D5 (DSLR). He also recently shot footage for the first ever 8K video from space using a RED Helium camera. This is available to view on YouTube. There are also numerous other cameras positioned around the space station to capture imagery on board.

13. **Where can I get pictures from the International Space Station in original quality?**
   All images posted to Alexander's social media channels throughout the Horizons mission are also available to view and download in high resolution on Flickr. If you choose to share these, please ensure you use the appropriate credits.

14. **Can you see stars and planets, and do they look different?**
   Stars and planets are visible from the International Space Station. In fact, Alexander has taken some stunning images of the Moon, Mars and Milky Way during his missions.

   From the perspective of the station, stars appear to be clearer and twinkle less than on Earth. This is because the station is positioned outside the atmosphere that causes twinkling and starlight doesn't have to travel through the same amount of gas and dust. The higher you are, the less of this atmosphere that starlight has to travel through.

   Planets also appear much brighter in space – particularly Jupiter, Venus and Mars.

15. **What happens during landing?**

    Astronauts return to Earth in the same vehicle they used to travel to the International Space Station. In the case of Alexander Gerst and his crew, this is the Soyuz MS-09.

    The Soyuz descent module must re-enter Earth's atmosphere at a specific angle to adequately reduce its speed. If the angle is too flat, the capsule could bounce back into space like a stone skipping across water. If the angle is too steep, the capsule could get too hot.

    The Soyuz's descent module has a heat shield that protects astronauts up to 1600 °C and astronauts experience a maximum gravitational load of 4 g when the capsule reaches an altitude of 35 km.

    In the unlikely event that the automatic control system fails, the crew members use a manual controller as backup. They train extensively to prepare for this possibility. Another option is the ballistic descent, in which the spacecraft starts spinning and flies a much steeper trajectory. The g load in this case will increase up to nine.

    When the capsule reaches an altitude of 10.5 km its speed has already decreased from 28,000 to 800 km/h. In order to decrease this speed even further, a series of parachutes are deployed. At a height of 8.5 km, the drogue shoot deploys the main parachute, which slows the capsule down to 22 km/h.

    Right before the capsule reaches the ground, soft lander thrusters fire. This further reduces its speed to approximately 5 km/h. Custom seats help absorb the shock of impact, but there really is no such thing as a soft landing in a Soyuz.

16. **Is it possible to build a space station that has artificial gravity?**

    In theory, this would be possible. Rotational simulated gravity has been proposed as a solution to the prolonged health effects of human spaceflight (perhaps most famously in *2001: A Space Odyssey*). However, the size and cost of producing this kind of spacecraft would be enormous. It would also negate one of the most important aspects of the International Space Station – the ability for researchers to conduct scientific experiments in microgravity.

17. **Why is research on the International Space Station important?**

    The International Space Station is the only laboratory we have that allows researchers to perform long-term experimentation in microgravity. It allows astronauts to run investigations that are not possible anywhere else, and the information we gather through this has many benefits back on Earth.

    European experiments on the International Space Station include research into the impact of gravity on muscle tone and time perception. The outcomes of time research could benefit the elderly, the incapacitated or those working in isolation, while research into muscle tone could benefit patients going through long periods of inactivity. Other experiments aim to build a better understanding of materials and their behaviors, interactions between atoms, and even artificial intelligence and robotics. All of this knowledge will be applied to the benefit of Earth.

18. What happens if someone on station gets sick?

Astronauts are subject to 2 weeks of quarantine before flying to space, and many precautions are taken to ensure the space station remains as hygienic as possible. All astronauts also receive 2–3 weeks of medical training in which they learn different skills such as how to pull a tooth, repair a filling or perform a cardiac massage.

Thankfully it has never happened before, but if any major medical intervention such as surgery is required there is a rescue capsule docked to the station ready to go at any moment. This is the vehicle in which the crew arrived.

The decision to return to Earth would not be made lightly, as it would put an end to the crew's mission, but this decision is not made by the astronauts alone. The International Space Station crew is supported at all times by an expert medical team on the ground.

19. What did Alexander Gerst study?

Alexander Gerst is a geophysicist and volcanologist by training. In 2003, he received a diploma in geophysics from the University of Karlsruhe, Germany, and a Master's degree in Earth sciences from the Victoria University of Wellington, New Zealand. Both degrees were awarded with distinction.

In 2010 he graduated with a doctorate in Natural Sciences from the Institute of Geophysics of the University of Hamburg, Germany. His dissertation was on geophysics and volcanic eruption dynamics.

20. How can I become an astronaut?

Becoming an astronaut is neither simple nor straightforward, as there are no schools for astronauts or university courses. So how do you become an astronaut and what qualifications and qualities do you need?

Space agencies look for the best people possible. Training an astronaut is a considerable investment for any agency. Training is lengthy and expensive, and the support needed both before and during a space mission is costly.

Astronauts need to be able to apply their considerable knowledge and skills to the tasks for which they have been trained, bear tremendous responsibility while in orbit and be determined to succeed.

A high level of education in scientific or technical disciplines, coupled with an outstanding professional background in research, application or education fields, possibly supported by the use of computer systems and applications, is essential. Previous experience with aircraft operations is a bonus, particularly if it involved responsible tasks such as being a test pilot or flight engineer.

Astronauts must undergo intense periods of training and may participate in spaceflights that last for months. During this time their body is subject to a great deal of stress, making good health and physical endurance essential.

They must also have an affinity for teamwork and adaptability, as they'll need to live with others in a small, closed and remote space. They must also possess a high degree of self-control and an equable temperament to cope with any stress or emergencies that may arise.

ESA is not currently recruiting for astronauts. However, there are many other avenues you can pursue to help progress Europe's role in space. From psychology and medicine to engineering, technology, science, architecture and communications, the options are almost endless. Working together we can go much further – to the Moon, Mars and beyond!

Index

© Springer Nature Switzerland AG 2020
J. O'Sullivan, *European Missions to the International Space Station*,
Springer Praxis Books, https://doi.org/10.1007/978-3-030-30326-6

319